Being Mortal

Being Mortal

Illness, Medicine and
What Matters in the End

Atul Gawande

PROFILE BOOKS | **wellcome collection**

First published in Great Britain in 2014 by
PROFILE BOOKS LTD
3A Exmouth House
Pine Street
London ECIR OJH
www.profilebooks.com

Published in association with Wellcome Collection

**wellcome
collection**

Wellcome Collection is a free visitor destination for the incurably curious. It explores the
connections between medicine, life and art in the past, present and future. Wellcome Collection
is part of the Wellcome Trust, a global charitable foundation dedicated to achieving
extraordinary improvements in human and animal health.

Wellcome Collection
183 Euston Road
London NW1 2BE
www.wellcomecollection.org

First published in the United States of America in 2014 by
Metropolitan Books, Henry Holt and Company, LLC

10 9 8

'Things Fall Apart' and 'Letting Go' were previously published in different form in the
New Yorker magazine.

Printed and bound in Great Britain by
Clays, Bungay, Suffolk

A CIP catalogue record for this book is available from the British Library.

ISBN 978 1 84668 581 1
Export ISBN 978 1 78125 394 6
eISBN 978 1 84765 786 2

The paper this book is printed on is certified by the © 1996 Forest Stewardship
Council A.C. (FSC). It is ancient-forest friendly. The printer holds FSC chain of
custody SGS-COC-2061

To Sara Bershtel

I see it now—this world is swiftly passing.

—the warrior Karna, in the *Mahabharata*

They come to rest at any kerb:
All streets in time are visited.

—Philip Larkin, "Ambulances"

Contents

Being Mortal

Introduction

I learned about a lot of things in medical school, but mortality wasn't one of them. Although I was given a dry, leathery corpse to dissect in my first term, that was solely a way to learn about human anatomy. Our textbooks had almost nothing on aging or frailty or dying. How the process unfolds, how people experience the end of their lives, and how it affects those around them seemed beside the point. The way we saw it, and the way our professors saw it, the purpose of medical schooling was to teach how to save lives, not how to tend to their demise.

The one time I remember discussing mortality was during an hour we spent on *The Death of Ivan Ilyich*, Tolstoy's classic novella. It was in a weekly seminar called Patient-Doctor—part of the school's effort to make us more rounded and humane physicians. Some weeks we would practice our physical examination etiquette; other weeks we'd learn about the effects of socioeconomics and race on health. And one afternoon we contemplated the suffering of Ivan Ilyich as he lay ill and worsening from some unnamed, untreatable disease.

In the story, Ivan Ilyich is forty-five years old, a midlevel Saint

Petersburg magistrate whose life revolves mostly around petty concerns of social status. One day, he falls off a stepladder and develops a pain in his side. Instead of abating, the pain gets worse, and he becomes unable to work. Formerly an "intelligent, polished, lively and agreeable man," he grows depressed and enfeebled. Friends and colleagues avoid him. His wife calls in a series of ever more expensive doctors. None of them can agree on a diagnosis, and the remedies they give him accomplish nothing. For Ilyich, it is all torture, and he simmers and rages at his situation.

"What tormented Ivan Ilyich most," Tolstoy writes, "was the deception, the lie, which for some reason they all accepted, that he was not dying but was simply ill, and he only need keep quiet and undergo a treatment and then something very good would result." Ivan Ilyich has flashes of hope that maybe things will turn around, but as he grows weaker and more emaciated he knows what is happening. He lives in mounting anguish and fear of death. But death is not a subject that his doctors, friends, or family can countenance. That is what causes him his most profound pain.

"No one pitied him as he wished to be pitied," writes Tolstoy. "At certain moments after prolonged suffering he wished most of all (though he would have been ashamed to confess it) for someone to pity him as a sick child is pitied. He longed to be petted and comforted. He knew he was an important functionary, that he had a beard turning grey, and that therefore what he longed for was impossible, but still he longed for it."

As we medical students saw it, the failure of those around Ivan Ilyich to offer comfort or to acknowledge what is happening to him was a failure of character and culture. The late-nineteenth-century Russia of Tolstoy's story seemed harsh and almost primitive to us. Just as we believed that modern medicine could probably have cured Ivan Ilyich of whatever disease he had, so too we took for granted that honesty and kindness were basic responsibilities of a modern doctor. We were confident that in such a situation we would act compassionately.

What worried us was knowledge. While we knew how to sympathize, we weren't at all certain we would know how to properly diagnose and treat. We paid our medical tuition to learn about the inner process of the body, the intricate mechanisms of its pathologies, and the vast trove of discoveries and technologies that have accumulated to stop them. We didn't imagine we needed to think about much else. So we put Ivan Ilyich out of our heads.

Yet within a few years, when I came to experience surgical training and practice, I encountered patients forced to confront the realities of decline and mortality, and it did not take long to realize how unready I was to help them.

I BEGAN WRITING when I was a junior surgical resident, and in one of my very first essays, I told the story of a man whom I called Joseph Lazaroff. He was a city administrator who'd lost his wife to lung cancer a few years earlier. Now, he was in his sixties and suffering from an incurable cancer himself—a widely metastatic prostate cancer. He had lost more than fifty pounds. His abdomen, scrotum, and legs had filled with fluid. One day, he woke up unable to move his right leg or control his bowels. He was admitted to the hospital, where I met him as an intern on the neurosurgical team. We found that the cancer had spread to his thoracic spine, where it was compressing his spinal cord. The cancer couldn't be cured, but we hoped it could be treated. Emergency radiation, however, failed to shrink the cancer, and so the neurosurgeon offered him two options: comfort care or surgery to remove the growing tumor mass from his spine. Lazaroff chose surgery. My job, as the intern on the neurosurgery service, was to get his written confirmation that he understood the risks of the operation and wished to proceed.

I'd stood outside his room, his chart in my damp hand, trying to figure out how to even broach the subject with him. The hope was that the operation would halt the progression of his spinal cord damage. It wouldn't cure him, or reverse his paralysis, or get him back to the life he had led. No matter what we did he had at most a few months to live, and the procedure was inherently dangerous. It required opening his chest, removing a rib, and collapsing a lung to get at his spine. Blood loss would be high. Recovery would be difficult. In his weakened state, he faced considerable risks of debilitating complications afterward. The operation posed a threat of both worsening and shortening his life. But the neurosurgeon had gone over these dangers, and Lazaroff had been clear that he wanted the operation. All I had to do was go in and take care of the paperwork.

Lying in his bed, Lazaroff looked gray and emaciated. I said that I was an intern and that I'd come to get his consent for surgery, which required confirming that he was aware of the risks. I said that the operation could remove the tumor but leave him with serious complications, such as paralysis or a stroke, and that it could even prove fatal. I tried to sound clear without being harsh, but my discussion put his back up. Likewise when his son, who was in the room, questioned whether heroic measures were a good idea. Lazaroff didn't like that at all.

"Don't you give up on me," he said. "You give me every chance I've got." Outside the room, after he signed the form, the son took me aside. His mother had died on a ventilator in intensive care, and at the time his father had said he did not want anything like that to happen to him. But now he was adamant about doing "everything."

I believed then that Mr. Lazaroff had chosen badly, and I still believe this. He chose badly not because of all the dangers but because the operation didn't stand a chance of giving him what he really wanted: his continence, his strength, the life he had previously known. He was pursuing little more than a fantasy at the risk of a prolonged and terrible death—which was precisely what he got.

The operation was a technical success. Over eight and a half hours, the surgical team removed the mass invading his spine and rebuilt the vertebral body with acrylic cement. The pressure on his spinal cord was gone. But he never recovered from the procedure. In intensive care, he developed respiratory failure, a systemic infection, blood clots from his immobility, then bleeding from the blood thinners to treat them. Each day we fell further behind. We finally had to admit he was dying. On the fourteenth day, his son told the team that we should stop.

It fell to me to take Lazaroff off the artificial ventilator that was keeping him alive. I checked to make sure that his morphine drip was turned up high, so he wouldn't suffer from air hunger. I leaned close and, in case he could hear me, said I was going to take the breathing tube out of his mouth. He coughed a couple of times when I pulled it out, opened his eyes briefly, and closed them. His breathing grew labored, then stopped. I put my stethoscope on his chest and heard his heart fade away.

Now, more than a decade after I first told Mr. Lazaroff's story, what strikes me most is not how bad his decision was but how much we all avoided talking honestly about the choice before him. We had no difficulty explaining the specific dangers of various treatment options, but we never really touched on the reality of his disease. His oncologists, radiation therapists, surgeons, and other doctors had all seen him through months of treatments for a problem that they knew could not be cured. We could never bring ourselves to discuss the larger truth about his condition or the ultimate limits of our capabilities, let alone what might matter most to him as he neared the end of his life. If he was pursuing a delusion, so were we. Here he was in the hospital, partially paralyzed from a cancer that had spread throughout his body. The chances that he could return to anything like the life he had even a few weeks earlier were zero. But admitting this and helping him cope with it seemed beyond us. We offered no acknowledgment or comfort or guidance. We just had another treatment he could undergo. Maybe something very good would result.

We did little better than Ivan Ilyich's primitive nineteenth-century doctors—worse, actually, given the new forms of physical torture we'd inflicted on our patient. It is enough to make you wonder, who are the primitive ones.

MODERN SCIENTIFIC CAPABILITY has profoundly altered the course of human life. People live longer and better than at any other time in history. But scientific advances have turned the processes of aging and dying into medical experiences, matters to be managed by health care professionals. And we in the medical world have proved alarmingly unprepared for it.

This reality has been largely hidden, as the final phases of life become less familiar to people. As recently as 1945, most deaths occurred in the home. By the 1980s, just 17 percent did. Those who somehow did die at home likely died too suddenly to make it to the hospital—say, from a massive heart attack, stroke, or violent injury—or were too isolated to get somewhere that could provide help. Across not just the United States but also the entire industrialized world, the experience of advanced aging and death has shifted to hospitals and nursing homes.

When I became a doctor, I crossed over to the other side of

the hospital doors and, although I had grown up with two doctors for parents, everything I saw was new to me. I had certainly never seen anyone die before and when I did it came as a shock. That wasn't because it made me think of my own mortality. Somehow the concept didn't occur to me, even when I saw people my own age die. I had a white coat on; they had a hospital gown. I couldn't quite picture it the other way round. I could, however, picture my family in their places. I'd seen multiple family members—my wife, my parents, and my children—go through serious, life-threatening illnesses. Even under dire circumstances, medicine had always pulled them through. The shock to me therefore was seeing medicine *not* pull people through. I knew theoretically that my patients could die, of course, but every actual instance seemed like a violation, as if the rules I thought we were playing by were broken. I don't know what game I thought this was, but in it we always won.

Dying and death confront every new doctor and nurse. The first times, some cry. Some shut down. Some hardly notice. When I saw my first deaths, I was too guarded to cry. But I dreamt about them. I had recurring nightmares in which I'd find my patients' corpses in my house—in my own bed.

"How did he get here?" I'd wonder in panic.

I knew I would be in huge trouble, maybe criminal trouble, if I didn't get the body back to the hospital without getting caught. I'd try to lift it into the back of my car, but it would be too heavy. Or I'd get it in, only to find blood seeping out like black oil until it overflowed the trunk. Or I'd actually get the corpse to the hospital and onto a gurney, and I'd push it down hall after hall, trying and failing to find the room where the person used to be. "Hey!" someone would shout and start chasing me. I'd wake up next to my wife in the dark, clammy and tachycardic. I felt that I'd killed these people. I'd failed.

Death, of course, is not a failure. Death is normal. Death may be the enemy, but it is also the natural order of things. I knew these truths abstractly, but I didn't know them concretely—that they could be truths not just for everyone but also for this person right in front of me, for this person I was responsible for.

The late surgeon Sherwin Nuland, in his classic book *How We Die,* lamented, "The necessity of nature's final victory was expected and accepted in generations before our own. Doctors were far more willing to recognize the signs of defeat and far less arrogant about denying them." But as I ride down the runway of the twenty-first century, trained in the deployment of our awesome arsenal of technology, I wonder exactly what being less arrogant really means.

You become a doctor for what you imagine to be the satisfaction of the work, and that turns out to be the satisfaction of competence. It is a deep satisfaction very much like the one that a carpenter experiences in restoring a fragile antique chest or that a science teacher experiences in bringing a fifth grader to that sudden, mind-shifting recognition of what atoms are. It comes partly from being helpful to others. But it also comes from being technically skilled and able to solve difficult, intricate problems. Your competence gives you a secure sense of identity. For a clinician, therefore, nothing is more threatening to who you think you are than a patient with a problem you cannot solve.

There's no escaping the tragedy of life, which is that we are all aging from the day we are born. One may even come to understand and accept this fact. My dead and dying patients don't haunt my dreams anymore. But that's not the same as saying one knows how to cope with what cannot be mended. I am in a profession that has succeeded because of its ability to fix. If your problem is fixable, we know just what to do. But if it's not? The fact that we have had no adequate answers to this question is troubling and has caused callousness, inhumanity, and extraordinary suffering.

This experiment of making mortality a medical experience is just decades old. It is young. And the evidence is it is failing.

THIS IS A book about the modern experience of mortality—about what it's like to be creatures who age and die, how medicine has changed the experience and how it hasn't, where our ideas about how to deal with our finitude have got the reality wrong. As I pass a decade in surgical practice and become middle-aged myself, I find that neither I nor my patients find our current state tolerable. But I have also found it unclear what the answers should be, or even whether any adequate ones are possible. I have the writer's and scientist's faith, however, that by pulling back the veil and peering in close, a person can make sense of what is most confusing or strange or disturbing.

You don't have to spend much time with the elderly or those with terminal illness to see how often medicine fails the people it is supposed to help. The waning days of our lives are given over to treatments that addle our brains and sap our bodies for a sliver's chance of benefit. They are spent in institutions—nursing homes and intensive care units—where regimented, anonymous routines cut us off from all the things that matter to us in life. Our reluctance to honestly examine the experience of aging and dying has increased the harm we inflict on people and denied them the basic comforts they most need. Lacking a coherent view of how people might live successfully all the way to their very end, we have allowed our fates to be controlled by the imperatives of medicine, technology, and strangers.

I wrote this book in the hope of understanding what has happened. Mortality can be a treacherous subject. Some will be alarmed by the prospect of a doctor's writing about the inevitability of decline and death. For many, such talk, however carefully framed, raises the specter of a society readying itself to sacrifice its sick and aged. But what if the sick and aged are *already* being sacrificed—victims of our refusal to accept the inexorability of our life cycle? And what if there are better approaches, right in front of our eyes, waiting to be recognized?

1 · *The Independent Self*

Growing up, I never witnessed serious illness or the difficulties of old age. My parents, both doctors, were fit and healthy. They were immigrants from India, raising me and my sister in the small college town of Athens, Ohio, so my grandparents were far away. The one elderly person I regularly encountered was a woman down the street who gave me piano lessons when I was in middle school. Later she got sick and had to move away, but it didn't occur to me to wonder where she went and what happened to her. The experience of a modern old age was entirely outside my perception.

In college, however, I began dating a girl in my dorm named Kathleen, and in 1985, on a Christmas visit to her home in Alexandria, Virginia, I met her grandmother Alice Hobson, who was seventy-seven at the time. She struck me as spirited and independent minded. She never tried to disguise her age. Her undyed white hair was brushed straight and parted on one side, Bette Davis–style. Her hands were speckled with age spots, and her skin was crinkled. She wore simple, neatly pressed blouses and dresses, a bit of lipstick, and heels long past when others would have considered it advisable.

As I came to learn over the years—for I would eventually marry Kathleen—Alice grew up in a rural Pennsylvania town known for its flower and mushroom farms. Her father was a flower farmer, growing carnations, marigolds, and dahlias, in acres of greenhouses. Alice and her siblings were the first members of their family to attend college. At the University of Delaware, Alice met Richmond Hobson, a civil engineering student. Thanks to the Great Depression, it wasn't until six years after their graduation that they could afford to get married. In the early years, Alice and Rich moved often for his work. They had two children, Jim, my future father-in-law, and then Chuck. Rich was hired by the Army Corps of Engineers and became an expert in large dam and bridge construction. A decade later, he was promoted to a job working with the corps's chief engineer at headquarters outside Washington, DC, where he remained for the rest of his career. He and Alice settled in Arlington. They bought a car, took road trips far and wide, and put away some money, too. They were able to upgrade to a bigger house and send their brainy kids off to college without need of loans.

Then, on a business trip to Seattle, Rich had a sudden heart attack. He'd had a history of angina and took nitroglycerin tablets to relieve the occasional bouts of chest pain, but this was 1965, and back then doctors didn't have much they could do about heart disease. He died in the hospital before Alice could get there. He was just sixty years old. Alice was fifty-six.

With her pension from the Army Corps of Engineers, she was able to keep her Arlington home. When I met her, she'd been living on her own in that house on Greencastle Street for twenty years. My in-laws, Jim and Nan, were nearby, but Alice lived completely independently. She mowed her own lawn and knew how to fix the plumbing. She went to the gym with her friend Polly. She liked to sew and knit and made clothes, scarves, and elaborate red-and-green Christmas stockings for everyone in the family, complete with a button-nosed Santa and their names across the top. She organized a group that took an annual subscription to attend performances at the Kennedy Center for the Performing Arts. She drove a big V8 Chevrolet Impala, sitting on a cushion to see over the dashboard. She ran errands, visited family, gave friends rides, and delivered meals-on-wheels for those with more frailties than herself.

As time went on, it became hard not to wonder how much longer she'd be able to manage. She was a petite woman, five feet tall at most, and although she bristled when anyone suggested it, she lost some height and strength with each passing year. When I married her granddaughter, Alice beamed and held me close and told me how happy the wedding made her, but she'd become too arthritic to share a dance with me. And still she remained in her home, managing on her own.

When my father met her, he was surprised to learn she lived by herself. He was a urologist, which meant he saw many elderly patients, and it always bothered him to find them living alone. The way he saw it, if they didn't already have serious needs, they were bound to develop them, and coming from India he felt it was the family's responsibility to take the aged in, give them company, and look after them. Since arriving in New York City in 1963 for his residency training, my father had embraced virtually every aspect of American culture. He gave up vegetarianism and discovered dating. He got a girlfriend, a pediatrics resident from a part of India where they didn't speak his language. When he married her, instead of letting my grandfather arrange his marriage, the family was scandalized. He became a tennis enthusiast, president of the local Rotary Club, and teller of bawdy jokes. One of his proudest days was July 4, 1976, the country's bicentennial, when he was made an American citizen in front of hundreds of cheering people in the grandstand at the Athens County Fair between the hog auction and the demolition derby. But one thing he could never get used to was how we treat our old and frail—leaving them to a life alone or isolating them in a series of anonymous facilities, their last conscious moments spent with nurses and doctors who barely knew their names. Nothing could have been more different from the world he had grown up in.

MY FATHER'S FATHER had the kind of traditional old age that, from a Western perspective, seems idyllic. Sitaram Gawande was a farmer in a village called Uti, some three hundred miles inland from Mumbai, where our ancestors had cultivated land for centuries. I remember visiting him with my parents and sister around the same time I met Alice, when he was more than a hundred years old. He was, by far, the oldest person I'd ever known. He walked with a cane, stooped like a bent stalk of wheat. He was so hard of hearing that people had to shout in his ear through a rubber tube. He was weak and sometimes needed help getting up from sitting. But he was a dignified man, with a tightly wrapped white turban, a pressed, brown argyle cardigan, and a pair of old-fashioned, thick-lensed, Malcolm X–style spectacles. He was surrounded and supported by family at all times, and he was revered—not in spite of his age but because of it. He was consulted on all important matters—marriages, land disputes, business decisions—and occupied a place of high honor in the family. When we ate, we served him first. When young people came into his home, they bowed and touched his feet in supplication.

In America, he would almost certainly have been placed in a nursing home. Health professionals have a formal classification system for the level of function a person has. If you cannot, without assistance, use the toilet, eat, dress, bathe, groom, get out of bed, get out of a chair, and walk—the eight "Activities of Daily Living"—then you lack the capacity for basic physical independence. If you cannot shop for yourself, prepare your own food, maintain your housekeeping, do your laundry, manage your medications, make phone calls, travel on your own, and handle your finances—the eight "Independent Activities of Daily Living"—then you lack the capacity to live safely on your own.

My grandfather could perform only some of the basic measures of independence, and few of the more complex ones. But in India, this was not of any dire consequence. His situation prompted no family crisis meeting, no anguished debates over what to do with him. It was clear that the family would ensure my grandfather could continue to live as he desired. One of my uncles and his family lived with him, and with a small herd of children, grandchildren, nieces, and nephews nearby, he never lacked for help.

The arrangement allowed him to maintain a way of life that few elderly people in modern societies can count on. The family made it possible, for instance, for him to continue to own and manage his farm, which he had built up from nothing—indeed, from worse than nothing. His father had lost all but two mortgaged acres and two emaciated bulls to a moneylender when the harvest failed one year. He then died, leaving Sitaram, his eldest son, with the debts. Just eighteen years old and newly married, Sitaram was forced to enter into indentured labor on the family's two remaining acres. At one point, the only food he and his bride could afford was bread and salt. They were starving to death. But he prayed and stayed at the plow, and his prayers were answered. The harvest was spectacular. He was able to not only put food on the table but also pay off his debts. In subsequent years, he expanded his two acres to more than two hundred. He became one of the richest landowners in the village and a moneylender himself. He had three wives, all of whom he outlived, and thirteen children. He emphasized education, hard work, frugality, earning your own way, staying true to your word, and holding others strictly accountable for doing the same. Throughout his life, he awoke before sunrise and did not go to bed until he'd done a nighttime inspection of every acre of his fields by horse. Even when he was a hundred he would insist on doing this. My uncles were worried he'd fall—he was weak and unsteady—but they knew it was important to him. So they got him a smaller horse and made sure that someone always accompanied him. He made the rounds of his fields right up to the year he died.

Had he lived in the West, this would have seemed absurd. It isn't safe, his doctor would say. If he persisted, then fell, and went to an emergency room with a broken hip, the hospital would not let him return home. They'd insist that he go to a nursing home. But in my grandfather's premodern world, how he wanted to live was his choice, and the family's role was to make it possible.

My grandfather finally died at the age of almost a hundred and ten. It happened after he hit his head falling off a bus. He was going to the courthouse in a nearby town on business, which itself seems crazy, but it was a priority to him. The bus began to move while he was getting off and, although he was accompanied by family, he fell. Most probably, he developed a subdural hematoma—bleeding inside his skull. My uncle got him home, and over the next couple of days he faded away. He got to live the way he wished and with his family around him right to the end.

FOR MOST OF human history, for those few people who actually survived to old age, Sitaram Gawande's experience was the norm. Elders were cared for in multigenerational systems, often with three generations living under one roof. Even when the nuclear family replaced the extended family (as it did in northern Europe several centuries ago), the elderly were not left to cope with the infirmities of age on their own. Children typically left home as soon as they were old enough to start families of their own. But one child usually remained, often the youngest daughter, if the parents survived into senescence. This was the lot of the poet Emily Dickinson, in Amherst, Massachusetts, in the mid-nineteenth century. Her elder brother left home, married, and started a family, but she and her younger sister stayed with their parents until they died. As it happened, Emily's father lived to the age of seventy-one, by which time she was in her forties, and her mother lived even longer. She and her sister ended up spending their entire lives in the parental home.

As different as Emily Dickinson's parents' life in America seems from that of Sitaram Gawande's in India, both relied on systems that shared the advantage of easily resolving the question of care for the elderly. There was no need to save up for a spot in a nursing home or arrange for meals-on-wheels. It was understood that parents would just keep living in their home, assisted by one or more of the children they'd raised. In contemporary societies, by contrast, old age and infirmity have gone from being a shared, multigenerational responsibility to a more or less private state—something experienced largely alone or with the aid of doctors and institutions. How did this happen? How did we go from Sitaram Gawande's life to Alice Hobson's?

One answer is that old age itself has changed. In the past,

surviving into old age was uncommon, and those who did survive served a special purpose as guardians of tradition, knowledge, and history. They tended to maintain their status and authority as heads of the household until death. In many societies, elders not only commanded respect and obedience but also led sacred rites and wielded political power. So much respect accrued to the elderly that people used to pretend to be older than they were, not younger, when giving their age. People have always lied about how old they are. Demographers call the phenomenon "age heaping" and have devised complex quantitative contortions to correct for all the lying in censuses. They have also noticed that, during the eighteenth century, in the United States and Europe, the direction of our lies changed. Whereas today people often understate their age to census takers, studies of past censuses have revealed that they used to overstate it. The dignity of old age was something to which everyone aspired.

But age no longer has the value of rarity. In America, in 1790, people aged sixty-five or older constituted less than 2 percent of the population; today, they are 14 percent. In Germany, Italy, and Japan, they exceed 20 percent. China is now the first country on earth with more than 100 million elderly people.

As for the exclusive hold that elders once had on knowledge and wisdom, that, too, has eroded, thanks to technologies of communication—starting with writing itself and extending to the Internet and beyond. New technology also creates new occupations and requires new expertise, which further undermines the value of long experience and seasoned judgment. At one time, we might have turned to an old-timer to explain the world. Now we consult Google, and if we have any trouble with the computer we ask a teenager.

Perhaps most important of all, increased longevity has brought about a shift in the relationship between the young and the old. Traditionally, surviving parents provided a source of much-needed stability, advice, and economic protection for young families seeking pathways to security. And because land-owners also tended to hold on to their property until death, the child who sacrificed everything to care for the parents could expect to inherit the whole homestead, or at least a larger portion than a child who moved away. But once parents were living markedly longer lives, tension emerged. For young people, the traditional family system became less a source of security than a struggle for control—over property, finances, and even the most basic decisions about how they could live.

And indeed, in my grandfather Sitaram's traditional household, generational tension was never far away. You can imagine how my uncles felt as their father turned a hundred and they entered old age themselves, still waiting to inherit land and gain economic independence. I learned of bitter battles in village families between elders and adult children over land and money. In the final year of my grandfather's life, an angry dispute erupted between him and my uncle with whom he lived. The original cause was unclear: perhaps my uncle had made a business decision without my grandfather; maybe my grandfather wanted to go out and no one in the family would go with him; maybe he liked to sleep with the window open and they liked to sleep with the window closed. Whatever the reason, the argument culminated (depending on who told the story) in Sitaram's either storming out of the house in the dead of night or being locked out. He somehow made it miles away to another relative's house and refused to return for two months.

Global economic development has changed opportunities for the young dramatically. The prosperity of whole countries depends on their willingness to escape the shackles of family expectation and follow their own path—to seek out jobs wherever they might be, do whatever work they want, marry whom they desire. So it was with my father's path from Uti to Athens, Ohio. He left the village first for university in Nagpur and then for professional opportunity in the States. As he became successful, he sent ever larger amounts of money home, helping to build new houses for his father and siblings, bring clean water and telephones to the village, and install irrigation systems that ensured harvests when the rainy seasons were bad. He even built a rural college nearby that he named for his mother. But there was no denying that he had left, and he wasn't going back.

Disturbed though my father was by the way America treated its elderly, the more traditional old age that my grandfather was able to maintain was possible only because my father's siblings had not left home as he had. We think, nostalgically, that we want the kind of old age my grandfather had. But the reason we do not have it is that, in the end, we do not actually want it. The historical pattern is clear: as soon as people got the resources and opportunity to abandon that way of life, they were gone.

THE FASCINATING THING is that, over time, it doesn't seem that the elderly have been especially sorry to see the children go. Historians find that the elderly of the industrial era did not suffer economically and were not unhappy to be left on their own. Instead, with growing economies, a shift in the pattern of property ownership occurred. As children departed home for opportunities elsewhere, parents who lived long lives found they could rent or even sell their land instead of handing it down. Rising incomes, and then pension systems, enabled more and more people to accumulate savings and property, allowing them to maintain economic control of their lives in old age and freeing them from the need to work until death or total disability. The radical concept of "retirement" started to take shape.

Life expectancy, which was under fifty in 1900, climbed to more than sixty by the 1930s, as improvements in nutrition, sanitation, and medical care took hold. Family sizes fell from an average of seven children in the mid-1800s to just over three after 1900. The average age at which a mother had her last child fell too—from menopause to thirty or younger. As a result, vastly more people lived to see their children reach adulthood. In the early twentieth century, a woman would have been fifty when her last child turned twenty-one, instead of in her sixties a century before. Parents had many years, easily a decade or more, before they or their children had to worry about old age.

So what they did was move on, just like their children. Given the opportunity, both parents and children saw separation as a form of freedom. Whenever the elderly have had the financial means, they have chosen what social scientists have called "intimacy at a distance." Whereas in early-twentieth-century America 60 percent of those over age sixty-five resided with a child, by the 1960s the proportion had dropped to 25 percent. By 1975 it was below 15 percent. The pattern is a worldwide one. Just 10 percent of Europeans over age eighty live with their children, and almost half live completely alone, without a spouse. In Asia, where the idea of an elderly parent being left to live alone has traditionally been regarded as shameful—the way my father saw it—the same radical shift is taking place. In China, Japan, and Korea, national statistics show the percentage of elderly living alone rising rapidly.

This is actually a sign of enormous progress. Choices for the elderly have proliferated. Del Webb, an Arizona real estate developer, popularized the term "retirement community" in 1960 when he launched Sun City, a community in Phoenix that was among the first to limit its residents to retirees. It was a controversial idea at the time. Most developers believed the elderly wanted more contact with other generations. Webb disagreed. He believed people in the last phase of their lives didn't want to live the way my grandfather did, with the family underfoot. He built Sun City as a place with an alternate vision of how people would spend what he called "their leisure years." It had a golf course, a shopping arcade, and a recreation center, and it offered the prospect of an active retirement of recreation and dining out with others like them to share it with. Webb's vision proved massively popular, and in Europe, the Americas, and even Asia, retirement communities have become a normal presence.

For those who had no interest in moving into such places—Alice Hobson, for instance—it became acceptable and feasible to remain in their own homes, living as they wanted to live, autonomously. That fact remains something to celebrate. There is arguably no better time in history to be old. The lines of power between the generations have been renegotiated, and not in the way it is sometimes believed. The aged did not lose status and control so much as share it. Modernization did not demote the elderly. It demoted the family. It gave people—the young and the old—a way of life with more liberty and control, including the liberty to be less beholden to other generations. The veneration of elders may be gone, but not because it has been replaced by veneration of youth. It's been replaced by veneration of the independent self.

THERE REMAINS ONE problem with this way of living. Our reverence for independence takes no account of the reality of what happens in life: sooner or later, independence will become impossible. Serious illness or infirmity will strike. It is as inevitable as sunset. And then a new question arises: If independence is what we live for, what do we do when it can no longer be sustained?

In 1992, Alice turned eighty-four. She was in striking health. She'd had to make a transition to false teeth and undergo removal of cataracts in both eyes. That was all. She'd had no major illnesses or hospitalizations. She still went to the gym with her friend Polly and did her own shopping and took care of her house. Jim and Nan offered her the option of turning their basement into an apartment for her. She might find it easier to be there, they said. She wouldn't hear of it. She had no intention of not living on her own.

But things began to change. On a mountain vacation with the family, Alice didn't turn up for lunch. She was found sitting in the wrong cabin, wondering where everyone was. We'd never seen her confused like that before. The family kept a close eye on her for the next few days, but nothing else untoward happened. We all let the matter drop.

Then Nan, visiting Alice at home one afternoon, noticed black-and-blue bruises up and down her leg. Had she fallen?

No, Alice said at first. But later she admitted that she'd taken a spill going down the wooden basement stairs. It was just a slip, she insisted. It could have happened to anyone. She'd be more careful next time.

Soon, however, she had more falls, several of them. No broken bones, but the family was getting worried. So Jim did what all families naturally do nowadays. He had her see a doctor.

The doctor did some tests. He found that she had thinning bones and recommended calcium. He fiddled with her medications and gave her some new prescriptions. But the truth was he didn't know what to do. We were not bringing him a fixable problem. Alice was unsteady. Her memory was slipping. The problems were only going to increase. Her independence would not be sustainable for long now. But he had no answers or direction or guidance. He could not even describe what to expect would happen.

2 · *Things Fall Apart*

Medicine and public health have transformed the trajectory of our lives. For all but our most recent history, death was a common, ever-present possibility. It didn't matter whether you were five or fifty. Every day was a roll of the dice. If you plotted the typical course of a person's health, it would look like this:

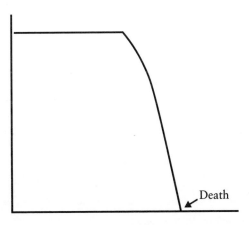

Life and health would putter along nicely, not a problem in the world. Then illness would hit and the bottom would drop out like a trap door—the way it did for my grandmother Gopikabai Gawande, who'd been perfectly well until the day she was struck by a fatal case of malaria, not even thirty years old, or for Rich Hobson, who had a heart attack on a business trip and then was gone.

Over the years, with medical progress, the bottom has tended to drop out later and later. The advent of sanitation and other public health measures sharply reduced the likelihood of death from infectious disease, especially in early childhood, and clinical advances dramatically reduced the mortality of childbirth and traumatic injuries. By the middle of the twentieth century, just four out of every hundred people in industrialized countries died before the age of thirty. And in the decades since, medicine found ways to cut the mortality of heart attacks, respiratory illnesses, stroke, and numerous other conditions that threaten in adult life. Eventually, of course, we all die of something. But even then, medicine has pushed the fatal moment of many diseases further outward. People with incurable cancers, for instance, can do remarkably well for a long time after diagnosis. They undergo treatment. Symptoms come under control. They resume regular life. They don't feel sick. But the disease, while slowed, continues progressing, like a night brigade taking out perimeter defenses. Eventually, it makes itself known, turning up in the lungs, or in the brain, or in the spine, as it did with Joseph Lazaroff. From there, the decline is often relatively rapid, much as in the past. Death occurs later, but the trajectory remains the same. In a matter of months or weeks, the body becomes overwhelmed. That is why, although the diagnosis may have been present for years, death can still come as a surprise. The road that seemed so straight and steady can still disappear, putting a person on a fast and steep slide down.

The pattern of decline has changed, however, for many chronic illnesses—emphysema, liver disease, and congestive heart failure, for example. Instead of just delaying the moment of the downward drop, our treatments can stretch the descent out until it ends up looking less like a cliff and more like a hilly road down the mountain:

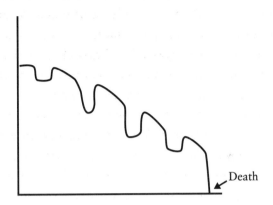

The road can have vertiginous drops but also long patches of recovered ground: we may not be able to stave off the damage, but we can stave off the death. We have drugs, fluids, surgery, intensive care units to get people through. They enter the hospital looking terrible, and some of what we do can make them look worse. But just when it looks like they've breathed their last, they rally. We make it possible for them to make it home—weaker and more impaired, though. They never return to their previous baseline. As illness progresses and organ damage worsens, a person becomes less able to withstand even minor problems. A simple cold can be fatal. The ultimate course is still downward until there finally comes a time when there is no recovery at all.

The trajectory that medical progress has made possible for many people, though, follows neither of these two patterns. Instead, increasingly large numbers of us get to live out a full life span and die of old age. Old age is not a diagnosis. There is always some final proximate cause that gets written down on the death certificate—respiratory failure, cardiac arrest. But in truth no single disease leads to the end; the culprit is just the accumulated crumbling of one's bodily systems while medicine carries out its maintenance measures and patch jobs. We reduce the blood pressure here, beat back the osteoporosis there, control this disease, track that one, replace a failed joint, valve, piston, watch the central processing unit gradually give out. The curve of life becomes a long, slow fade:

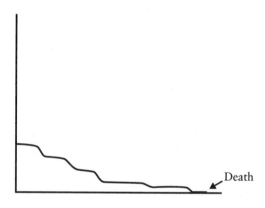

The progress of medicine and public health has been an incredible boon—people get to live longer, healthier, more productive lives than ever before. Yet traveling along these altered paths, we regard living in the downhill stretches with a kind of embarrassment. We need help, often for long periods of time, and regard that as a weakness rather than as the new normal and expected state of affairs. We're always trotting out some story of a ninety-seven-year-old who runs marathons, as if such cases were not miracles of biological luck but reasonable expectations for all. Then, when our bodies fail to live up to this fantasy, we feel as if we somehow have something to apologize for.

Those of us in medicine don't help, for we often regard the patient on the downhill as uninteresting unless he or she has a discrete problem we can fix. In a sense, the advances of modern medicine have given us two revolutions: we've undergone a biological transformation of the course of our lives and also a cultural transformation of how we think about that course.

THE STORY OF aging is the story of our parts. Consider the teeth. The hardest substance in the human body is the white enamel of the teeth. With age, it nonetheless wears away, allowing the softer, darker layers underneath to show through. Meanwhile, the blood supply to the pulp and the roots of the teeth atrophies, and the flow of saliva diminishes; the gums tend to become inflamed and pull away from the teeth, exposing the base, making them unstable and elongating their appearance, especially the lower ones. Experts say they can gauge a person's age to within five years from the examination of a single tooth—if the person has any teeth left to examine.

Scrupulous dental care can help avert tooth loss, but growing old gets in the way. Arthritis, tremors, and small strokes, for example, make it difficult to brush and floss, and because nerves become less sensitive with age, people may not realize that they have cavity and gum problems until it's too late. In the course of a normal lifetime, the muscles of the jaw lose about 40 percent of their mass and the bones of the mandible lose about 20 percent, becoming porous and weak. The ability to chew declines, and people shift to softer foods, which are generally higher in fermentable carbohydrates and more likely to cause cavities. By the age of sixty, people in an industrialized country like the United States have lost, on average, a third of their teeth. After eighty-five, almost 40 percent have no teeth at all.

Even as our bones and teeth soften, the rest of our body hardens. Blood vessels, joints, the muscle and valves of the heart, and even the lungs pick up substantial deposits of calcium and turn stiff. Under a microscope, the vessels and soft tissues display the same form of calcium that you find in bone. When you reach inside an elderly patient during surgery, the aorta and other major vessels can feel crunchy under your fingers. Research has found that loss of bone density may be an even better predictor of death from atherosclerotic disease than cholesterol levels. As we age, it's as if the calcium seeps out of our skeletons and into our tissues.

To maintain the same volume of blood flow through our narrowed and stiffened blood vessels, the heart has to generate increased pressure. As a result, more than half of us develop hypertension by the age of sixty-five. The heart becomes thicker-walled from having to pump against the pressure, and less able to respond to the demands of exertion. The peak output of the heart therefore decreases steadily from the age of thirty. People become gradually less able to run as far or as fast as they used to or to climb a flight of stairs without becoming short of breath.

As the heart muscle thickens, muscle elsewhere thins. Around age forty, one begins to lose muscle mass and power. By age eighty, one has lost between a quarter and a half of one's muscle weight.

You can see all these processes play out just in the hand: 40 percent of the muscle mass of the hand is in the thenar muscles, the muscles of the thumb, and if you look carefully at the palm of an older person, at the base of the thumb, you will notice that the musculature is not bulging but flat. In a plain X-ray, you will see speckles of calcification in the arteries and translucency of the bones, which, from age fifty, lose their density at a rate of nearly 1 percent per year. The hand has twenty-nine joints, each of which is prone to destruction from osteoarthritis, and this will give the joint surfaces a ragged, worn appearance. The joint space collapses. You can see bone touching bone. What the person feels is swelling around the joints, reduced range of motion of the wrist, diminished grip, and pain. The hand also has forty-eight named nerve branches. Deterioration of the cutaneous mechanoreceptors in the pads of the fingers produces loss of sensitivity to touch. Loss of motor neurons produces loss of dexterity. Handwriting degrades. Hand speed and vibration sense decline. Using a standard mobile phone, with its tiny buttons and touch screen display, becomes increasingly unmanageable.

This is normal. Although the processes can be slowed—diet and physical activity can make a difference—they cannot be stopped. Our functional lung capacity decreases. Our bowels slow down. Our glands stop functioning. Even our brains shrink: at the age of thirty, the brain is a three-pound organ that barely fits inside the skull; by our seventies, gray-matter loss leaves almost an inch of spare room. That's why elderly people like my grandfather are so much more prone to cerebral bleeding after a blow to the head—the brain actually rattles around inside. The earliest portions to shrink are generally the frontal lobes, which govern judgment and planning, and the hippocampus, where memory is organized. As a consequence, memory and the ability to gather and weigh multiple ideas—to multitask—peaks in midlife and then gradually declines. Processing speeds start decreasing well before age forty (which may be why mathematicians and physicists commonly do their best work in their youth). By age eighty-five, working memory and judgment are sufficiently impaired that 40 percent of us have textbook dementia.

WHY WE AGE is the subject of vigorous debate. The classical view is that aging happens because of random wear and tear.

The newest view holds that aging is more orderly and genetically programmed. Proponents of this view point out that animals of similar species and exposure to wear and tear have markedly different life spans. The Canada goose has a longevity of 23.5 years; the emperor goose only 6.3 years. Perhaps animals are like plants, with lives that are, to a large extent, internally governed. Certain species of bamboo, for instance, form a dense stand that grows and flourishes for a hundred years, flowers all at once, and then dies.

The idea that living things shut down instead of wearing down has received substantial support in recent years. Researchers working with the now famous worm C. *elegans* (twice in one decade, Nobel Prizes went to scientists doing work on the little nematode) were able, by altering a single gene, to produce worms that live more than twice as long and age more slowly. Scientists have since come up with single-gene alterations that increase the life spans of fruit flies, mice, and yeast.

These findings notwithstanding, the preponderance of the evidence is against the idea that our life spans are programmed into us. Remember that for most of our hundred-thousand-year existence—all but the past couple of hundred years—the average life span of human beings has been thirty years or less. (Research suggests that subjects of the Roman Empire had an average life expectancy of twenty-eight years.) The natural course was to die before old age. Indeed, for most of history, death was a risk at every age of life and had no obvious connection with aging, at all. As Montaigne wrote, observing late-sixteenth-century life, "To die of age is a rare, singular, and extraordinary death, and so much less natural than others: it is the last and extremest kind of dying." So today, with our average life span in much of the world climbing past eighty years, we are already oddities living well beyond our appointed time. When we study aging what we are trying to understand is not so much a natural process as an unnatural one.

It turns out that inheritance has surprisingly little influence on longevity. James Vaupel, of the Max Planck Institute for Demographic Research, in Rostock, Germany, notes that only 3 percent of how long you'll live, compared with the average, is explained by your parents' longevity; by contrast, up to 90 percent of how tall you are is explained by your parents' height. Even genetically identical twins vary widely in life span: the typical gap is more than fifteen years.

If our genes explain less than we imagined, the classical wear-and-tear model may explain more than we knew. Leonid Gavrilov, a researcher at the University of Chicago, argues that human beings fail the way all complex systems fail: randomly and gradually. As engineers have long recognized, simple devices typically do not age. They function reliably until a critical component fails, and the whole thing dies in an instant. A windup toy, for example, works smoothly until a gear rusts or a spring breaks, and then it doesn't work at all. But complex systems—power plants, say—have to survive and function despite having thousands of critical, potentially fragile components. Engineers therefore design these machines with multiple layers of redundancy: with backup systems, and backup systems for the backup systems. The backups may not be as efficient as the first-line components, but they allow the machine to keep going even as damage accumulates. Gavrilov argues that, within the parameters established by our genes, that's exactly how human beings appear to work. We have an extra kidney, an extra lung, an extra gonad, extra teeth. The DNA in our cells is frequently damaged under routine conditions, but our cells have a number of DNA repair systems. If a key gene is permanently damaged, there are usually extra copies of the gene nearby. And, if the entire cell dies, other cells can fill in.

Nonetheless, as the defects in a complex system increase, the time comes when just one more defect is enough to impair the whole, resulting in the condition known as frailty. It happens to power plants, cars, and large organizations. And it happens to us: eventually, one too many joints are damaged, one too many arteries calcify. There are no more backups. We wear down until we can't wear down anymore.

It happens in a bewildering array of ways. Hair grows gray, for instance, simply because we run out of the pigment cells that give hair its color. The natural life cycle of the scalp's pigment cells is just a few years. We rely on stem cells under the surface to migrate in and replace them. Gradually, however, the stem-cell reservoir is used up. By the age of fifty, as a result, half of the average person's hairs have gone gray.

Inside skin cells, the mechanisms that clear out waste products slowly break down and the residue coalesces into a clot of gooey yellow-brown pigment known as lipofuscin. These are the age spots we see in skin. When lipofuscin accumulates in sweat glands, the sweat glands cannot function, which helps explain why we become so susceptible to heat stroke and heat exhaustion in old age.

The eyes go for different reasons. The lens is made of crystallin proteins that are tremendously durable, but they change chemically in ways that diminish their elasticity over time—hence the farsightedness that most people develop beginning in their fourth decade. The process also gradually yellows the lens. Even without cataracts (the whitish clouding of the lens that occurs with age, excessive ultraviolet exposure, high cholesterol, diabetes, and cigarette smoking), the amount of light reaching the retina of a healthy sixty-year-old is one-third that of a twenty-year-old.

I spoke to Felix Silverstone, who for twenty-four years was the senior geriatrician at the Parker Jewish Institute, in New York, and who has published more than a hundred studies on aging. There is, he told me, "no single, common cellular mechanism to the aging process." Our bodies accumulate lipofuscin and oxygen free-radical damage and random DNA mutations and numerous other microcellular problems. The process is gradual and unrelenting.

I asked Silverstone whether gerontologists have discerned any particular, reproducible pathway to aging. "No," he said. "We just fall apart."

THIS IS NOT, to say the least, an appealing prospect. People naturally prefer to avoid the subject of their decrepitude. There have been dozens of bestselling books on aging, but they tend to have titles such as *Younger Next Year*, *The Fountain of Age*, *Ageless*, or—my favorite—*The Sexy Years*. Still, there are costs to averting our eyes from the realities. We put off dealing with the adaptations that we need to make as a society. And we blind ourselves to the opportunities that exist to change the individual experience of aging for the better.

As medical progress has extended our lives, the result has been what's called the "rectangularization" of survival. Throughout most of human history, a society's population formed a sort of pyramid: young children represented the largest portion—the base—and each successively older cohort represented a smaller and smaller group. In 1950, children under the age of five were 11 percent of the US population, adults aged forty-five to forty-nine were 6 percent, and those over eighty were 1 percent. Today, we have as many fifty-year-olds as five-year-olds. In thirty years, there will be as many people over eighty as there are under five. The same pattern is emerging throughout the industrialized world.

Few societies have come to grips with the new demography. We cling to the notion of retirement at sixty-five—a reasonable notion when those over sixty-five were a tiny percentage of the population but increasingly untenable as they approach 20 percent. People are putting aside less in savings for old age now than they have at any time since the Great Depression. More than half of the very old now live without a spouse and we have fewer children than ever before, yet we give virtually no thought to how we will live out our later years alone.

Equally worrying, and far less recognized, medicine has been slow to confront the very changes that it has been responsible for—or to apply the knowledge we have about how to make old age better. Although the elderly population is growing rapidly, the number of certified geriatricians the medical profession has put in practice has actually fallen in the United States by 25 percent between 1996 and 2010. Applications to training programs in adult primary care medicine have plummeted, while fields like plastic surgery and radiology receive applications in record numbers. Partly, this has to do with money—incomes in geriatrics and adult primary care are among the lowest in medicine. And partly, whether we admit it or not, a lot of doctors don't like taking care of the elderly.

"Mainstream doctors are turned off by geriatrics, and that's because they do not have the faculties to cope with the Old Crock," Felix Silverstone, the geriatrician, explained to me. "The Old Crock is deaf. The Old Crock has poor vision. The Old Crock's memory might be somewhat impaired. With the Old Crock, you have to slow down, because he asks you to repeat what you are saying or asking. And the Old Crock doesn't just have a chief complaint—the Old Crock has fifteen chief complaints. How in the world are you going to cope with all of them? You're overwhelmed. Besides, he's had a number of these things for fifty years or so. You're not going to cure something he's had for fifty years. He has high blood pressure. He has diabetes. He has arthritis. There's nothing glamorous about taking care of any of those things."

There is, however, a skill to it, a developed body of professional expertise. One may not be able to fix such problems, but one can manage them. And until I visited my hospital's geriatrics clinic and saw the work that the clinicians there do, I did not fully grasp the nature of the expertise involved, or how important it could be for all of us.

THE GERIATRICS CLINIC—OR, as my hospital calls it, the Center for Older Adult Health (even in a clinic geared to people eighty years or older, patients view words like "geriatrics" or just "elderly" askance)—is only one floor below my surgery clinic. I passed by it almost every day for years, and I can't remember ever giving it a moment's thought. One morning, however, I wandered downstairs and, with the permission of the patients, sat in on a few visits with Juergen Bludau, the chief geriatrician.

"What brings you here today?" the doctor asked Jean Gavrilles, his first patient of the morning. She was eighty-five years old, with short, frizzy white hair, oval glasses, a lavender knit shirt, and a sweet, ready smile. Small but sturdy in appearance, she had come in walking steadily, her purse and coat clutched under one arm, her daughter trailing behind her, no support required beyond her mauve orthopedic shoes. She said that her internist had recommended that she come.

About anything in particular? the doctor asked.

The answer, it seemed, was yes and no. The first thing she mentioned was a lower-back pain that she'd had for months, which shot down her leg and sometimes made it difficult to get out of bed or up from a chair. She also had bad arthritis, and she showed us her fingers, which were swollen at the knuckles and bent out to the sides with what's called a swan-neck deformity. She'd had both knees replaced a decade earlier. She had high blood pressure, "from stress," she said, before handing Bludau her list of medications. She had glaucoma and needed to have eye exams every four months. She never used to have "bathroom problems," but lately, she admitted, she'd started wearing a pad. She'd also had surgery for colon cancer and, by the way, she now had a lung nodule that the radiology report said could be a metastasis—a biopsy was recommended.

Bludau asked her to tell him about her life, and it reminded me of the life Alice lived when I first met her at my in-laws'. Gavrilles said that she lived alone, except for her Yorkshire terrier, in a single-family house in the West Roxbury section of Boston. Her husband died of lung cancer twenty-three years ago. She did not drive. She had a son living in the area who did her shopping once a week and checked on her each day—"just to see if I'm still alive," she joked. Another son and two daughters lived farther away, but they helped as well. Otherwise, she took care of herself quite capably. She did her own cooking and cleaning. She managed her medicines and her bills.

"I have a system," she said.

She had a high school education, and during World War II she'd worked as a riveter at the Charlestown Navy Yard. She also worked for a time at the Jordan Marsh department store in downtown Boston. But that was a long time ago. She stuck to home now, with her yard and her terrier and her family when they visited.

The doctor asked her about her day in great detail. She usually woke around five or six o'clock, she said—she didn't seem to need much sleep anymore. She would get out of bed as the back pain allowed, take a shower, and get dressed. Downstairs, she'd take her medicines, feed the dog, and eat breakfast. Bludau asked what she had for breakfast that day. Cereal and a banana, she said. She hated bananas, but she'd heard they were good for her potassium, so she was afraid to stop. After breakfast, she'd take her dog for a little walk in the yard. She did chores—laundry, cleaning, and the like. In the late morning, she took a break to watch *The Price Is Right*. At lunchtime, she had a sandwich and orange juice. If the weather was nice, she'd sit out in the yard afterward. She'd loved working in her garden, but she could no longer do that. The afternoons were slow. She might do some more chores. She might nap or talk on the phone. Eventually, she would make dinner—a salad and maybe a baked potato or a scrambled egg. At night, she watched the Red Sox or the Patriots or college basketball—she loved sports. She usually went to bed at about midnight.

Bludau asked her to sit on the examining table. As she struggled to climb up, her balance teetering on the step, the doctor held her arm. He checked her blood pressure, which was normal. He examined her eyes and ears and had her open her mouth. He listened to her heart and lungs briskly, with his stethoscope. He began to slow down only when he looked at her hands. The nails were neatly trimmed.

"Who cuts your nails?" he asked.

"I do," Gavrilles replied.

I tried to think what could be accomplished in this visit. She was in good condition for her age, but she faced everything from advancing arthritis and incontinence to what might be metastatic colon cancer. It seemed to me that, with just a forty-minute visit, Bludau needed to triage by zeroing in on either the most potentially life-threatening problem (the possible metastasis) or the problem that bothered her the most (the back pain). But this was evidently not what he thought. He asked almost nothing about either issue. Instead, he spent much of the exam looking at her feet.

"Is that really necessary?" she asked, when he instructed her to take off her shoes and socks.

"Yes," he said. After she'd left, he told me, "You must always examine the feet." He described a bow-tied gentleman who seemed dapper and fit, until his feet revealed the truth: he couldn't bend down to reach them, and they turned out not to have been cleaned in weeks, suggesting neglect and real danger.

Gavrilles had difficulty taking her shoes off, and, after watching her struggle a bit, Bludau leaned in to help. When he got her socks off, he took her feet in his hands, one at a time. He inspected them inch by inch—the soles, the toes, the web spaces. Then he helped her get her socks and shoes back on and gave her and her daughter his assessment.

She was doing impressively well, he said. She was mentally sharp and physically strong. The danger for her was losing what she had. The single most serious threat she faced was not the lung nodule or the back pain. It was falling. Each year, about 350,000 Americans fall and break a hip. Of those, 40 percent end up in a nursing home, and 20 percent are never able to walk again. The three primary risk factors for falling are poor balance, taking more than four prescription medications, and muscle weakness. Elderly people without these risk factors have a 12 percent chance of falling in a year. Those with all three risk factors have almost a 100 percent chance. Jean Gavrilles had at least two. Her balance was poor. Though she didn't need a walker, he had noticed her splay-footed gait as she came in. Her feet were swollen. The toenails were unclipped. There were sores between the toes. And the balls of her feet had thick, rounded calluses.

She was also on five medications. Each was undoubtedly useful, but together the usual side effects would include dizziness. In addition, one of the blood pressure medications was a diuretic, and she seemed to drink few liquids, risking dehydration and a worsening of the dizziness. Her tongue was bone-dry when Bludau examined it.

She did not have significant muscle weakness, and that was good. When she got out of her chair, he said, he noted that she had not used her arms to push herself up. She simply stood up—a sign of well-preserved muscle strength. From the details of the day she described, however, she did not seem to be eating nearly enough calories to maintain that strength. Bludau asked her whether her weight had changed recently. She admitted that she had lost about seven pounds in the previous six months.

The job of any doctor, Bludau later told me, is to support quality of life, by which he meant two things: as much freedom from the ravages of disease as possible and the retention of enough function for active engagement in the world. Most doctors treat disease and figure that the rest will take care of itself. And if it doesn't—if a patient is becoming infirm and heading toward a nursing home—well, that isn't really a *medical* problem, is it?

To a geriatrician, though, it *is* a medical problem. People can't stop the aging of their bodies and minds, but there are ways to make it more manageable and to avert at least some of the worst effects. So Bludau referred Gavrilles to a podiatrist, whom he wanted her to visit once every four weeks, for better care of her feet. He didn't see medications that he could eliminate, but he switched her diuretic to a blood pressure medicine that wouldn't cause dehydration. He recommended that she eat a snack during the day, get all the low-calorie and low-cholesterol food out of the house, and see whether family or friends could join her for more meals. "Eating alone is not very stimulating," he said. And he asked her to see him again in three months, so that he could make sure the plan was working.

Almost a year later, I checked in with Gavrilles and her daughter. She'd turned eighty-six. She was eating better and had even gained a pound or two. She still lived comfortably and independently in her own home. And she had not had a single fall.

ALICE BEGAN FALLING long before I met Juergen Bludau or Jean Gavrilles and grasped the possibilities that might have been. Neither I nor anyone else in the family understood that her falls were a loud alarm bell or that a few simple changes might have preserved, for at least some time longer, her independence and the life she wanted. Her doctors never understood this either. Matters just kept getting worse.

Next came not a fall but a car accident. Backing her Chevy Impala out of her driveway, she shot across the street, over the curb, and through a yard, and could not stop the car until it ended up in some bushes against her neighbor's house. The family speculated that she'd stomped on the accelerator instead of the brake. Alice insisted the accelerator had got stuck. She thought of herself as a good driver and hated the idea that anyone would think that the problem was her age.

The body's decline creeps like a vine. Day to day, the changes can be imperceptible. You adapt. Then something happens that finally makes it clear that things are no longer the same. The falls didn't do it. The car accident didn't do it. Instead, it was a scam that did.

Not long after the car accident, Alice hired two men to perform tree and yard work. They set a reasonable price with her but clearly saw her as a mark. When they finished the job, they told her that she owed nearly a thousand dollars. She balked. She was very careful and organized about money. But they got angry and threatening, and, cornered, she wrote the check. She was shaken but also embarrassed and told no one about it, hoping she could put it behind her. A day later, the men returned late in the evening and demanded she pay more. She argued with them, but in the end she wrote that check, too. The ultimate total was more than seven thousand dollars. Again, she wasn't going to say anything. Neighbors, however, heard the raised voices at Alice's doorstep and called the police.

The men were gone by the time the police arrived. A policeman took a statement from Alice and promised to investigate further. She still didn't want to tell the family about what had happened. But she knew this was trouble and after a while finally told my father-in-law, Jim.

He spoke to the neighbors who'd reported the crime. They mentioned that they had become worried for her. She no longer seemed safe living on her own. There was this incident and the Impala in the bushes. There was also what they observed of how difficult managing matters as ordinary as getting her trash to the curb had become.

The police caught the scam artists and arrested them for grand larceny. The men were convicted and sentenced to prison, which should have been satisfying for Alice. But instead the whole process kept the events, and the reminders of her growing vulnerability, alive and lingering when she would have dearly loved to have set them behind her.

Soon after the scammers were caught, Jim suggested that he and Alice go together to look at retirement homes. It was just to see what they were like, he said. But they both knew where this was going.

DECLINE REMAINS OUR fate; death will someday come. But until that last backup system inside each of us fails, medical care can influence whether the path is steep and precipitate or more gradual, allowing longer preservation of the abilities that matter most in your life. Most of us in medicine don't think about this. We're good at addressing specific, individual problems: colon cancer, high blood pressure, arthritic knees. Give us a disease, and we can do something about it. But give us an elderly woman with high blood pressure, arthritic knees, and various other ailments besides—an elderly woman at risk of losing the life she enjoys—and we hardly know what to do and often only make matters worse.

Several years ago, researchers at the University of Minnesota identified 568 men and women over the age of seventy who were living independently but were at high risk of becoming disabled because of chronic health problems, recent illness, or cognitive changes. With their permission, the researchers randomly assigned half of them to see a team of geriatric nurses and doctors—a team dedicated to the art and science of managing old age. The others were asked to see their usual physician, who was notified of their high-risk status. Within eighteen months, 10 percent of the patients in both groups had died. But the patients who had seen a geriatrics team were a quarter less likely to become disabled and half as likely to develop depression. They were 40 percent less likely to require home health services.

These were stunning results. If scientists came up with a device—call it an automatic defrailer—that wouldn't extend your life but would slash the likelihood you'd end up in a nursing home or miserable with depression, we'd be clamoring for it. We wouldn't care if doctors had to open up your chest and plug the thing into your heart. We'd have pink-ribbon campaigns to get one for every person over seventy-five. Congress would be holding hearings demanding to know why forty-year-olds couldn't get them installed. Medical students would be jockeying to become defrailulation specialists, and Wall Street would be bidding up company stock prices.

Instead, it was just geriatrics. The geriatric teams weren't doing lung biopsies or back surgery or insertion of automatic defrailers. What they did was to simplify medications. They saw that arthritis was controlled. They made sure toenails were trimmed and meals were square. They looked for worrisome signs of isolation and had a social worker check that the patient's home was safe.

How do we reward this kind of work? Chad Boult, the geriatrician who was the lead investigator of the University of Minnesota study, can tell you. A few months after he published the results, demonstrating how much better people's lives were with specialized geriatric care, the university closed the division of geriatrics.

"The university said that it simply could not sustain the financial losses," Boult said from Baltimore, where he had moved to join the Johns Hopkins Bloomberg School of Public Health. On average, in Boult's study, the geriatric services cost the hospital $1,350 more per person than the savings they produced, and Medicare, the insurer for the elderly, does not cover that cost. It's a strange double standard. No one insists that a $25,000 pacemaker or a coronary-artery stent save money for insurers. It just has to *maybe* do people some good. Meanwhile, the twenty-plus members of the proven geriatrics team at the University of Minnesota had to find new jobs. Scores of medical centers across the country have shrunk or closed their geriatrics units. Many of Boult's colleagues no longer advertise their geriatric training for fear that they'll get too many elderly patients. "Economically, it has become too difficult," Boult said.

But the dismal finances of geriatrics are only a symptom of a deeper reality: people have not insisted on a change in priorities. We all like new medical gizmos and demand that policy makers ensure they are paid for. We want doctors who promise to fix things. But geriatricians? Who clamors for geriatricians? What geriatricians do—bolster our resilience in old age, our capacity to weather what comes—is both difficult and unappealingly limited. It requires attention to the body and its alterations. It requires vigilance over nutrition, medications, and living situations. And it requires each of us to contemplate the unfixables in our life, the decline we will unavoidably face, in order to make the small changes necessary to reshape it. When the prevailing fantasy is that we can be ageless, the geriatrician's uncomfortable demand is that we accept we are not.

FOR FELIX SILVERSTONE, managing aging and its distressing realities was the work of a lifetime. He was a national leader in geriatrics for five decades. But when I met him he was himself eighty-seven years old. He could feel his own mind and body wearing down, and much of what he spent his career studying was no longer at a remove from him.

Felix had been fortunate. He didn't have to stop working, even after he suffered a heart attack in his sixties that cost him half his heart function; nor was he stopped by a near cardiac arrest at the age of seventy-nine.

"One evening, sitting at home, I suddenly became aware of palpitations," he told me. "I was just reading, and a few minutes later I became short of breath. A little bit after that, I began to feel heavy in the chest. I took my pulse, and it was over two hundred."

He is the sort of person who, in the midst of chest pain, would take the opportunity to examine his own pulse.

"My wife and I had a little discussion about whether or not to call an ambulance. We decided to call."

When Felix got to the hospital, the doctors had to shock him to bring his heart back. He'd had ventricular tachycardia, and an automatic defibrillator was implanted in his chest. Within a few weeks, he felt well again, and his doctor cleared him to return to work full time. He stayed in medical practice after the attack, multiple hernia repairs, gallbladder surgery, arthritis that all but ended his avid piano playing, compression fractures of his aging spine that stole three full inches of his five-foot-seven-inch height, and hearing loss.

"I switched to an electronic stethoscope," he said. "They're a nuisance, but they're very good."

Finally, at eighty-two, he had to retire. The problem wasn't his health; it was that of his wife, Bella. They'd been married for more than sixty years. Felix had met Bella when he was an intern and she was a dietitian at Kings County Hospital, in Brooklyn. They brought up two sons in Flatbush. When the boys left home, Bella got her teaching certificate and began working with children who had learning disabilities. In her seventies, however, retinal disease diminished her vision, and she had to stop working. A decade later, she'd become almost completely blind. Felix no longer felt safe leaving her at home alone, and in 2001 he gave up his practice. They moved to Orchard Cove, a retirement community in Canton, Massachusetts, outside Boston, where they could be closer to their sons.

"I didn't think I would survive the change," Felix said. He'd observed in his patients how difficult the transitions of age were. Examining his last patient, packing up his home, he felt that he was about to die. "I was taking apart my life as well as the house," he recalled. "It was terrible."

We were sitting in a library off Orchard Cove's main lobby. There was light streaming through a picture window, tasteful art on the walls, white upholstered Federal-style armchairs. It was like a nice hotel, only with no one under seventy-five walking around. Felix and Bella had a two-bedroom apartment with forest views and plenty of space. In the living room, Felix had a grand piano and, at his desk, piles of medical journals that he still subscribed to—"for my soul," he said. Theirs was an independent-living unit. It came with housekeeping, linen changes, and dinner each evening. When they needed to, they could upgrade to assisted living, which provides three prepared meals and up to an hour with a personal-care assistant each day.

This was not the average retirement community, but even in an average one rent runs $32,000 a year. Entry fees are typically $60,000 to $120,000 on top of that. Meanwhile, the median income of people eighty and older is only about $15,000. More than half of the elderly living in long-term-care facilities run through their entire savings and have to go on government assistance—welfare—in order to afford it. Ultimately, the average American spends a year or more of old age disabled and living in a nursing home (at more than five times the yearly cost of independent living), which is a destination Felix was desperately hoping to avoid.

He was trying to note the changes he experienced objectively, like the geriatrician he is. He noticed that his skin had dried out. His sense of smell was diminished. His night vision had become poor, and he tired easily. He had begun to lose teeth. But he took what measures he could. He used lotion to avoid skin cracks; he protected himself from the heat; he got on an exercise bike three times a week; he saw a dentist twice a year.

He was most concerned about the changes in his brain. "I can't think as clearly as I used to," he said. "I used to be able to read the *New York Times* in half an hour. Now it takes me an hour and a half." Even then, he wasn't sure that he understood as much as he did before, and his memory gave him trouble. "If I go back and look at what I've read, I recognize that I went through it, but sometimes I don't really remember it," he said. "It's a matter of short-term registration. It's hard to get the signal in and have it stay put."

He made use of methods that he once taught his patients. "I try to deliberately focus on what I'm doing, rather than do it automatically," he told me. "I haven't lost the automaticity of action, but I can't rely on it the way I used to. For example, I can't think about something else and get dressed and be sure I've gotten all the way dressed." He recognized that the strategy of trying to be more deliberate didn't always work, and he sometimes told me the same story twice in a conversation. The lines of thought in his mind would fall into well-worn grooves and, however hard he tried to put them onto a new path, sometimes they resisted. Felix's knowledge as a geriatrician forced him to recognize his decline, but it didn't make it easier to accept.

"I get blue occasionally," he said. "I think I have recurring episodes of depression. They are not enough to disable me, but they are . . ." He paused to find the right word. "They are uncomfortable."

What buoyed him, despite his limitations, was having a purpose. It was the same purpose, he said, that sustained him in medicine: to be of service, in some way, to those around him. He had been in Orchard Cove for only a few months before he was helping to steer a committee to improve the health care services there. He formed a journal-reading club for retired physicians. He even guided a young geriatrician through her first independent research study—a survey of the residents' attitudes toward Do Not Resuscitate orders.

More important was the responsibility that he felt for his children and grandchildren—and most of all for Bella. Her blindness and memory troubles had made her deeply dependent. Without him, she would have been in a nursing home. He helped her dress and administered her medicines. He made her breakfast and lunch. He took her on walks and to doctor's appointments. "She is my purpose now," he said.

Bella didn't always like his way of doing things.

"We argue constantly—we're at each other about a lot of things," Felix said. "But we're also very forgiving."

He did not feel this responsibility to be a burden. With the narrowing of his own life, his ability to look after Bella had become his main source of self-worth.

"I am exclusively her caregiver," he said. "I am glad to be." And this role had heightened his sense that he must be attentive to the changes in his own capabilities; he would be no good to her if he wasn't honest with himself about his own limitations.

One evening, Felix invited me to dinner. The formal dining hall was restaurant-like, with reserved seating, table service, and jackets required. I was wearing my white hospital coat and had to borrow a navy blazer from the maître d' in order to be seated. Felix, in a brown suit and a stone-colored oxford shirt, gave his arm to Bella, who wore a blue-flowered knee-length dress that he'd picked out for her, and guided her to the table. She was amiable and chatty and had youthful-seeming eyes. But once she'd been seated, she couldn't find the plate in front of her, let alone the menu. Felix ordered for her: wild-rice soup, an omelette, mashed potatoes, and mashed cauliflower. "No salt," he instructed the waiter; she had high blood pressure. He ordered salmon and mashed potatoes for himself. I had the soup and a London broil.

When the food arrived, Felix told Bella where she could find the different items on her plate by the hands of a clock. He put a fork in her hand. Then he turned to his own meal.

Both made a point of chewing slowly. She was the first to choke. It was the omelette. Her eyes watered. She began to cough. Felix guided her water glass to her mouth. She took a drink and managed to get the omelette down.

"As you get older, the lordosis of your spine tips your head forward," he said to me. "So when you look straight ahead it's like looking up at the ceiling for anyone else. Try to swallow while looking up: you'll choke once in a while. The problem is common in the elderly. Listen." I realized that I could hear someone in the dining room choking on his food every minute or so. Felix turned to Bella. "You have to eat looking down, sweetie," he said.

A couple of bites later, though, he himself was choking. It was the salmon. He began coughing. He turned red. Finally, he was able to cough up the bite. It took a minute for him to catch his breath.

"Didn't follow my own advice," he said.

Felix Silverstone was, without question, up against the debilities of his years. Once, it would have been remarkable simply to have lived to see eighty-seven. Now the remarkable thing was the control he'd maintained over his life. When he started in geriatric practice, it was almost inconceivable that an eighty-seven-year-old with his history of health problems could live independently, care for his disabled wife, and continue to contribute to research.

Partly, he had been lucky. His memory, for example, had not deteriorated badly. But he had also managed his old age well. His goal has been modest: to have as decent a life as medical knowledge and the limits of his body would allow. So he saved and did not retire early and was therefore not in financial straits. He kept his social contacts and avoided isolation. He monitored his bones and teeth and weight. And he made sure to find a doctor who had the geriatric skills to help him hold on to an independent life.

I ASKED CHAD Boult, the geriatrics professor, what could be done to ensure that there are enough geriatricians for the surging elderly population. "Nothing," he said. "It's too late." Creating geriatric specialists takes time, and we already have far too few. In a year, fewer than three hundred doctors will complete geriatrics training in the United States, not nearly enough to replace the geriatricians going into retirement, let alone meet the needs of the next decade. Geriatric psychiatrists, nurses, and social workers are equally needed, and in no better supply. The situation in countries outside the United States appears to be little different. In many, it is worse.

Yet Boult believes that we still have time for another strategy: he would direct geriatricians toward training all primary care doctors and nurses in caring for the very old, instead of providing the care themselves. Even this is a tall order—97 percent of medical students take no course in geriatrics, and the strategy requires that the nation pay geriatric specialists to teach rather than to provide patient care. But if the will is there, Boult estimates that it would be possible to establish courses in every medical school, nursing school, school of social work, and internal-medicine training program within a decade.

"We've got to do something," he said. "Life for older people can be better than it is today."

"I CAN STILL drive, you know," Felix Silverstone said to me after our dinner together. "I'm a very good driver."

He had to run an errand to refill Bella's prescriptions in Stoughton, a few miles away, and I asked if I could come along. He had a ten-year-old gold Toyota Camry with automatic transmission and 39,000 miles on the odometer. It was pristine, inside and out. He backed out of a narrow parking space and zipped out of the garage. His hands did not shake. Taking the streets of Canton at dusk on a new-moon night, he brought the car to an even stop at the red lights, signaled when he was supposed to, took turns without a hitch.

I was, I admit, braced for disaster. The risk of a fatal car crash with a driver who's eighty-five or older is more than three times higher than it is with a teenage driver. The very old are the highest-risk drivers on the road. I thought of Alice's wreck and considered how lucky she was that no child had been in her neighbor's yard. A few months earlier, in Los Angeles, George Weller was convicted of manslaughter after he confused the accelerator with the brake pedal and plowed his Buick into a crowd of shoppers at the Santa Monica Farmers Market. Ten people were killed, and more than sixty were injured. He was eighty-six.

But Felix showed no difficulties. At one point during our drive, poorly marked road construction at an intersection channeled our line of cars almost directly into oncoming traffic. Felix corrected course swiftly, pulling over into the proper lane. There was no saying how much longer he would be able to count on his driving ability. Someday, the hour would come when he would have to give up his keys.

At that moment, though, he wasn't concerned; he was glad simply to be on the road. The evening traffic was thin as he turned onto Route 138. He brought the Camry to a tick over the 45-mile-per-hour speed limit. He had his window rolled down and his elbow on the sash. The air was clear and cool, and we listened to the sound of the wheels on the pavement.

"The night is lovely, isn't it?" he said.

3 · *Dependence*

It is not death that the very old tell me they fear. It is what happens short of death—losing their hearing, their memory, their best friends, their way of life. As Felix put it to me, "Old age is a continuous series of losses." Philip Roth put it more bitterly in his novel *Everyman*: "Old age is not a battle. Old age is a massacre."

With luck and fastidiousness—eating well, exercising, keeping our blood pressure under control, getting medical help when we need it—people can often live and manage a very long time. But eventually the losses accumulate to the point where life's daily requirements become more than we can physically or mentally manage on our own. As fewer of us are struck dead out of the blue, most of us will spend significant periods of our lives too reduced and debilitated to live independently.

We do not like to think about this eventuality. As a result, most of us are unprepared for it. We rarely pay more than glancing attention to how we will live when we need help until it's too late to do much about it.

When Felix came to this crossroads, the orthopedic shoe to drop wasn't his. It was Bella's. Year by year, I witnessed the progression in her difficulties. Felix remained in astonishingly good health right into his nineties. He had no medical crises and maintained his weekly exercise regimen. He continued to teach chaplaincy students about geriatrics and to serve on Orchard Cove's health committee. He didn't even have to stop driving. But Bella was fading. She lost her vision completely. Her hearing became poor. Her memory became markedly impaired. When we had dinner, she had to be reminded more than once that I was sitting across from her.

She and Felix felt the sorrows of their losses but also the pleasures of what they still had. Although she might not have been able to remember me or others she didn't know too well, she enjoyed company and conversation and sought both out. Moreover, she and Felix still had their own, private, decades-long conversation that had never stopped. He found great purpose in caring for her, and she, likewise, found great meaning in being there for him. The physical presence of each other gave them comfort. He dressed her, bathed her, helped feed her. When they walked, they held hands. At night, they lay in bed in each other's arms, awake and nestling for a while, before finally drifting off to sleep. Those moments, Felix said, remained among their most cherished. He felt they knew each other, and loved each other, more than at any time in their nearly seventy years together.

One day, however, they had an experience that revealed just how fragile their life had become. Bella developed a cold, causing fluid to accumulate in her ears. An eardrum ruptured. And with that she became totally deaf. That was all it took to sever the thread between them. With her blindness and memory problems, the hearing loss made it impossible for Felix to achieve any kind of communication with her. He tried drawing out letters on the palm of her hand but she couldn't make them out. Even the simplest matters—getting her dressed, for instance—became a nightmare of confusion for her. Without sensory grounding, she lost track of time of day. She grew severely confused, at times delusional and agitated. He couldn't take care of her. He became exhausted from stress and lack of sleep.

He didn't know what to do, but there was a system for such situations. The people at the residence proposed transferring her to a skilled nursing unit—a nursing home floor. He couldn't bear the thought of it. No, he said. She needed to stay at home with him.

Before the issue was forced, they got a reprieve. Two and a half weeks into the ordeal, Bella's right eardrum mended and, although the hearing in her left ear was lost permanently, the hearing in her right ear came back.

"Our communication is more difficult," Felix said. "But at least it is possible."

I asked what he would do if the hearing in her right ear went again or if there were some other such catastrophe, and he told me he didn't know. "I'm in dread of what would happen if she becomes too hard for me to care for," he said. "I try not to think too far ahead. I don't think about next year. It's too depressing. I just think about next week."

It's the route people the world over take, and that is understandable. But it tends to backfire. Eventually, the crisis they dreaded arrived. They were walking together when, suddenly, Bella fell. He wasn't sure what had happened. They'd been walking slowly. The ground was flat. He'd had her by the arm. But she went down in a heap and snapped the fibula in both her legs—the long, thin outer bone that runs from knee to ankle. The emergency room doctors had to cast each of her limbs to above the knee. What Felix feared most had happened. Her needs became massively more than he could handle. Bella was forced to move to the nursing home floor, where she could have round-the-clock aides and nurses looking after her.

You might think that this would have been a relief for both Bella and Felix, lifting all kinds of burdens of physical care from them. But the experience was more complicated than that. On the one hand, the staff members were nothing but professional. They took over most of the tasks Felix had long managed so laboriously—the bathing, toileting, dressing, and all the other routine needs of a person who has become severely disabled. They freed him to spend his time as he wished, whether with Bella or on his own. But for all the staff members' efforts, Felix and Bella could find their presence exasperating. Some tended to Bella more as a patient than as a person. She had a certain way she liked her hair brushed, for instance, but no one asked or figured it out. Felix had worked out the best method to cut up her food so she could swallow it without difficulty, how to position her so she was most comfortable, how to dress her the way she preferred. But no matter how much he tried to show the staff, many of them did not see the point. Sometimes, in exasperation, he'd give up and simply redo whatever they had done, causing conflict and resentment.

"We were getting in each other's way," Felix said.

He worried too that the unfamiliar surroundings were making Bella confused. After a few days, he decided to move her back home. He'd just have to figure out how to deal with her.

Their apartment was only a floor away. But somehow that made all the difference. Exactly why can be hard to pinpoint. Felix still ended up hiring an around-the-clock staff of nurses and aides. And the remaining six weeks until the casts could come off were physically exhausting for him. Yet he was relieved.

He and Bella felt more control over her life. She was in her own place, in her own bed, with him beside her. And that mattered tremendously to him. Because four days after the casts came off, four days after she'd begun walking again, she died.

They'd sat down to lunch. She turned to him and said, "I don't feel well." Then she collapsed. An ambulance whisked her to the local hospital. He didn't want to slow the medics down. So he let them go and followed after in his car. She died in the short time between her arrival and his.

When I saw him three months later, he was still despondent. "I feel as if a part of my body is missing. I feel as if I have been dismembered," he told me. His voice cracked and his eyes were rimmed red. He had one great solace, however: that she hadn't suffered, that she'd got to spend her last few weeks in peace at home in the warmth of their long love, instead of up on a nursing floor, a lost and disoriented patient.

ALICE HOBSON HAD something very much like the same dread of leaving her home. It was the one place where she felt she belonged and remained in charge of her life. But after the incident with the men who had victimized her, it was apparent that she wasn't safe living on her own anymore. My father-in-law organized a few visits to senior living residences for her. "She didn't care for this process," Jim said, but she reconciled herself to it. He was determined to find a place she would like and thrive in. But it was not to be. As I watched the aftermath, I gradually began to understand the reasons why—and they were reasons that bring into question our entire system of care for the dependent and debilitated.

Jim looked for a place that was within a reasonable driving distance for the family and within a price range she could afford with the proceeds of selling her house. He also wanted a community that offered a "continuum of care"—much like Orchard Cove, where I visited Felix and Bella—with apartments for independent living and a floor with the around-the-clock nursing capabilities that she might someday need. He came up with a variety of places for them to visit—nearer ones and farther ones, for-profit and not-for-profit.

The place Alice ultimately chose was a high-rise senior-living complex that I will call Longwood House, a nonprofit facility affiliated with the Episcopal Church. Some of her friends from church lived there. The drive to and from Jim's home was barely ten minutes. The community was active and thriving. To Alice and the family, it had by far the greatest appeal.

"Most of the others were too commercial," Jim said.

She moved in during the fall of 1992. Her one-bedroom independent-living apartment was more spacious than I'd expected. It had a full kitchen, enough room for her dining set, and plenty of light. My mother-in law, Nan, made sure it got a fresh coat of paint and arranged for a decorator Alice had used before to help place furniture and hang pictures.

"It means something when you can move in and see all your things in their own places—your own silver in your kitchen drawer," Nan said.

But when I saw Alice a few weeks after her move, she didn't seem at all happy or adjusted. Never one to complain, she didn't say anything angry or sad or bitter, but she was withdrawn in a way I hadn't seen before. She remained recognizably herself, but the light had gone out from behind her eyes.

At first I thought that this had to do with the loss of her car and the freedom that came with it. When she moved into Longwood House, she'd brought her Chevy Impala and fully intended to keep driving. But on her very first day there, when she went to take the car out for some errands, it was gone. She called the police and reported it stolen. An officer arrived, took a description, and promised an investigation. A while later, Jim arrived, and, on a hunch, looked in the Giant Food store parking lot next door. There it was. She had got confused and parked in the wrong lot without realizing it. Mortified, she gave up driving for good. In one day, she lost her car as well as her home.

But there seemed to be more to her sense of loss and unhappiness. She had a kitchen but stopped cooking. She took her meals in the Longwood House dining room with everyone else but ate little, lost weight, and didn't seem to like having the company. She avoided organized group activities, even the ones she might have enjoyed—a sewing circle like the one she'd had at her church, a book group, gym and fitness classes, trips to the Kennedy Center. The community offered opportunities to organize activities of your own if you didn't like what was on offer. But she stuck to herself. We thought she was depressed. Jim and Nan took her to see a doctor, who put her on medication. It didn't help. Somewhere along the seven-mile drive between the house she'd given up on Greencastle Street and Longwood House, her life fundamentally changed in ways she did not want but could do nothing about.

THE IDEA OF being unhappy in a place as comfortable as Longwood House would have seemed laughable at one time. In 1913, Mabel Nassau, a Columbia University graduate student, conducted a neighborhood study of the living conditions of one hundred elderly people in Greenwich Village—sixty-five women and thirty-five men. In this era before pensions and Social Security, all were poor. Only twenty-seven were able to support themselves—living off savings, taking in lodgers, or doing odd jobs like selling newspapers, cleaning homes, mending umbrellas. Most were too ill or debilitated to work.

One woman, for instance, whom Nassau called Mrs. C., was a sixty-two-year-old widow who'd made just enough as a domestic servant to afford a small back room with an oil stove in a rooming house. Illness had recently ended her work, however, and she now had severe leg swelling with varicose veins that left her bedbound. Miss S. was "unusually sick" and had a seventy-two-year-old brother with diabetes who, in this era before insulin treatment, was fast becoming crippled and emaciated as the disease killed him. Mr. M. was a sixty-seven-year-old Irish former longshoreman who'd been left disabled by a paralytic stroke. A large number had become simply "feeble," by which Nassau seemed to mean that they were too senile to manage for themselves.

Unless family could take such people in, they had virtually no options left except a poorhouse, or almshouse, as it was often called. These institutions went back centuries in Europe and the United States. If you were elderly and in need of help but did not have a child or independent wealth to fall back on, a poorhouse was your only source of shelter. Poorhouses were grim, odious places to be incarcerated—and that was the telling term used at the time. They housed poor of all types—elderly paupers, out-of-luck immigrants, young drunks, the mentally ill—and their function was to put the "inmates" to work for their presumed intemperance and moral turpitude. Supervisors usually treated elderly paupers leniently in work assignments, but they were inmates like the rest. Husbands and wives were separated. Basic physical care was lacking. Filth and dilapidation were the norm.

A 1912 report from the Illinois State Charities Commission described one county's poorhouse as "unfit to decently house animals." The men and women lived without any attempt at classification by age or needs in bare ten-by-twelve-foot rooms infested with bedbugs. "Rats and mice overrun the place. . . . Flies swarm [the] food. . . . There are no bathtubs." A 1909 Virginia report described elderly people dying untended, receiving inadequate nutrition and care, and contracting tuberculosis from uncontrolled contagion. Funds were chronically inadequate for disabled care. In one case, the report noted, a warden, faced with a woman who tended to wander off and no staff to mind her, made her carry a twenty-eight-pound ball and chain.

Nothing provoked greater terror for the aged than the prospect of such institutions. Nonetheless, by the 1920s and 1930s, when Alice and Richmond Hobson were young, two-thirds of poorhouse residents were elderly. Gilded Age prosperity had sparked embarrassment about these conditions. Then the Great Depression sparked a nationwide protest movement. Elderly middle-class people who'd worked and saved all their lives found their savings wiped out. In 1935, with the passage of Social Security, the United States joined Europe in creating a system of national pensions. Suddenly a widow's future was secure, and retirement, once the exclusive provenance of the rich, became a mass phenomenon.

In time, poorhouses passed from memory in the industrialized world, but they persist elsewhere. In developing countries, they have become common, because economic growth is breaking up the extended family without yet producing the affluence to protect the elderly from poverty and neglect. In India, I have noticed that the existence of such places is often unacknowledged, but on a recent visit to New Delhi I readily found examples. Their appearance seemed straight out of Dickens—or those old state reports.

The Guru Vishram Vridh ashram, for instance, is a charity-run old age home in a slum on the south edge of New Delhi, where open sewage ran in the streets and emaciated dogs rummaged in piles of trash. The home is a converted warehouse—a vast, open room with scores of disabled elderly people on cots and floor mattresses pushed up against one another like a large sheet of postage stamps. The proprietor, G. P. Bhagat, who appeared to be in his forties, was clean-cut and professional looking, with a cell phone that rang every two minutes. He said he'd been called by God to open the place eight years before and subsisted on donations. He said he never turned anyone away as long as he had an open bed. About half of the residents were deposited there by retirement homes and hospitals if they couldn't pay their bills. The other half were found in the streets and parks by volunteers or the police. All suffered from a combination of debility and poverty.

The place had more than a hundred people when I visited. The youngest was sixty and the oldest past a century. Those on the first floor had only "moderate" needs. Among them, I met a Sikh man crawling awkwardly along the ground, in a squat, like a slow-moving frog—hands-feet, hands-feet, hands-feet. He said he used to own an electrical shop in an upscale section of New Delhi. His daughter became an accountant, his son a software engineer. Two years ago something happened to him—he described chest pain and what sounded like a series of strokes. He spent two and a half months in the hospital, paralyzed. The bills rose. His family stopped visiting. Eventually the hospital dropped him off here. Bhagat said he sent a message to the family through the police saying the man would like to come home. They denied knowing him.

Up a narrow staircase was the second-floor ward for patients with dementia and other severe disabilities. An old man stood by a wall wailing out-of-tune songs at the top of his lungs. Next to him a woman with white, cataractal eyes muttered to herself.

Several staff members worked their way through the residents, feeding them and keeping them clean the best they could. The din and the smell of urine were overpowering. I tried to talk to a couple of the residents through my translator, but they were too confused to answer questions. A deaf and blind woman lying on a mattress nearby was shouting a few words over and over again. I asked the translator what she was saying. The translator shook her head—the words made no sense—and then she bolted down the stairs. It was too much for her. It was as close to a vision of hell as I've ever experienced.

"These people are on the last stage of their journey," Bhagat said, looking out upon the mass of bodies. "But I can't provide the kind of facility they really require."

In the course of Alice's lifetime, the industrialized world's elderly have escaped the threat of such a fate. Prosperity has enabled even the poor to expect nursing homes with square meals, professional health services, physical therapy, and bingo. They've eased debility and old age for millions and made proper care and safety a norm to an extent that the inmates of poorhouses could not imagine. Yet still, most consider modern old age homes frightening, desolate, even odious places to spend the last phase of one's life. We need and desire something more.

LONGWOOD HOUSE SEEMINGLY had everything going for it. The facility was up to date, with top ratings for safety and care. Alice's quarters enabled her to have the comforts of her old home in a safer, more manageable situation. The arrangements were tremendously reassuring for her children and extended family. But they weren't for Alice. She never got used to being there or accepted it. No matter what the staff or our family did for her, she grew only more miserable.

I asked her about this. But she couldn't put her finger on what made her unhappy. The most common complaint she made is one I've heard often from nursing home residents I've met: "It just isn't home." To Alice, Longwood House was a mere facsimile of home. And having a place that genuinely feels like your home can seem as essential to a person as water to a fish.

A few years ago, I read about the case of Harry Truman, an eighty-three-year-old man who, in March 1980, refused to budge from his home at the foot of Mount Saint Helens near Olympia, Washington, when the volcano began to steam and rumble. A former World War I pilot and Prohibition-era bootlegger, he'd owned his lodge on Spirit Lake for more than half a century. Five years earlier, he'd been widowed. So now it was just him and his sixteen cats on his fifty-four acres of property beneath the mountain. Three years earlier, he'd fallen off the lodge roof shoveling snow and broken his leg. The doctor told him he was "a damn fool" to be working up there at his age.

"Damn it!" Truman shot back. "I'm eighty years old and at eighty, I have the right to make up my mind and do what I want to do."

As eruption threatened, the authorities told everyone living in the vicinity to clear out. But Truman wasn't going anywhere. For more than two months, the volcano smoldered. Authorities extended the evacuation zone to ten miles around the mountain. Truman stubbornly remained. He didn't believe the scientists, with their uncertain and sometimes conflicting reports. He worried his lodge would be looted and vandalized, as another lodge on Spirit Lake was. And regardless, this home was his life.

"If this place is gonna go, I want to go with it," he said. " 'Cause if I lost it, it would kill me in a week anyway." He attracted reporters with his straight-talking, curmudgeonly way, holding forth with a green John Deere cap on his head and a tall glass of bourbon and Coke in his hand. The local police thought about arresting him for his own good but decided not to, given his age and the bad publicity they'd have to endure. They offered to bring him out every chance they got. He steadfastly refused. He told a friend, "If I die tomorrow, I've had a damn good life. I've done everything I could do, and I've done everything I ever wanted to do."

The blast came at 8:40 a.m. on May 18, 1980, with the force of an atomic bomb. The entire lake disappeared under the massive lava flow, burying Truman and his cats and his home with it. In the aftermath, he became an icon—the old man who had stayed in his house, taken his chances, and lived life on his own terms in an era when that possibility seemed to have all but disappeared. The people of nearby Castlerock constructed a memorial to him at the town's entrance that still stands, and there was a television movie starring Art Carney.

Alice wasn't facing a volcano, but she might as well have been. Giving up her home on Greencastle Street meant giving up the life she had built for herself over decades. The things that made Longwood House so much safer and more manageable than the house were precisely what made it hard for her to endure. Her apartment might have been called "independent living," but it involved the imposition of more structure and supervision than she'd ever had to deal with before. Aides watched her diet. Nurses monitored her health. They observed her growing unsteadiness and made her use a walker. This was reassuring for Alice's children, but she didn't like being nannied or controlled. And the regulation of her life only increased with time. When the staff became concerned that she was missing doses of her medications, they informed her that unless she kept her medications with the nurses and came down to their station twice a day to take them under direct supervision, she would have to move out of independent living to the nursing home wing. Jim and Nan hired a part-time aide named Mary to help Alice comply, to give her some company, and to stave off the day she would have to transfer. She liked Mary. But having her hanging around the apartment for hours on end, often with little to do, only made the situation more depressing.

For Alice, it must have felt as if she had crossed into an alien land that she would never be allowed to leave. The border guards were friendly and cheerful enough. They promised her a nice place to live where she'd be well taken care of. But she didn't really want anyone to take care of her; she just wanted to live a life of her own. And those cheerful border guards had taken her keys and her passport. With her home went her control.

People saw Harry Truman as a hero. There was never going to be a Longwood House for Harry Truman of Spirit Lake, and Alice Hobson of Arlington, Virginia, didn't want there to be one for her either.

HOW DID WE wind up in a world where the only choices for the very old seem to be either going down with the volcano or yielding all control over our lives? To understand what happened, you have to trace the story of how we replaced the poorhouse with the kinds of places we have today—and it turns out to be a medical story. Our old age homes didn't develop out of a desire to give the frail elderly better lives than they'd had in those dismal places. We didn't look around and say to ourselves, "You know, there's this phase of people's lives in which they can't really cope on their own, and we ought to find a way to make it manageable." No, instead we said, "This looks like a medical problem. Let's put these people in the hospital. Maybe the doctors can figure something out." The modern nursing home developed from there, more or less by accident.

In the middle part of the twentieth century, medicine was undergoing a rapid and historic transformation. Before that time, if you fell seriously ill, doctors usually tended to you in your own bed. The function of hospitals was mainly custodial. As the great physician-writer Lewis Thomas observed, describing his internship at Boston City Hospital in 1937, "If being in a hospital bed made a difference, it was mostly the difference produced by warmth, shelter, and food, and attentive, friendly care, and the matchless skill of the nurses in providing these things. Whether you survived or not depended on the natural history of the disease itself. Medicine made little or no difference."

From World War II onward, the picture shifted radically. Sulfa, penicillin, and then numerous other antibiotics became available for treating infections. Drugs to control blood pressure and treat hormonal imbalances were discovered. Breakthroughs in everything from heart surgery to artificial respirators to kidney transplantation became commonplace. Doctors became heroes, and the hospital transformed from a symbol of sickness and despondency to a place of hope and cure.

Communities could not build hospitals fast enough. In America, in 1946, Congress passed the Hill-Burton Act, which provided massive amounts of government funds for hospital construction. Two decades later the program had financed more than nine thousand new medical facilities across the country. For the first time, most people had a hospital nearby, and this became true across the industrialized world.

The magnitude of this transformation is impossible to overstate. For most of our species' existence, people were fundamentally on their own with the sufferings of their body. They depended on nature and chance and the ministry of family and religion. Medicine was just another a tool you could try, no different from a healing ritual or a family remedy and no more effective. But as medicine became more powerful, the modern hospital brought a different idea. Here was a place where you could go saying, "Cure me." You checked in and gave over every part of your life to doctors and nurses: what you wore, what you ate, what went into the different parts of your body and when. It wasn't always pleasant, but, for a rapidly expanding range of problems, it produced unprecedented results. Hospitals learned how to eliminate infections, remove cancerous tumors, reconstruct shattered bones. They could fix hernias and heart valves and hemorrhaging stomach ulcers. They became the normal place for people to go with their bodily troubles, including the elderly.

Meanwhile, policy planners had assumed that establishing a pension system would end poorhouses, but the problem did not go away. In America, in the years following the passage of the Social Security Act of 1935, the number of elderly in poorhouses refused to drop. States moved to close them but found they could not. The reason old people wound up in poorhouses, it turned out, was not just that they didn't have money to pay for a home. They were there because they'd become too frail, sick, feeble, senile, or broken down to take care of themselves anymore, and they had nowhere else to turn for help. Pensions provided a way of allowing the elderly to manage independently as long as possible in their retirement years. But pensions hadn't provided a plan for that final, infirm stage of mortal life.

As hospitals sprang up, they became a comparatively more attractive place to put the infirm. That was finally what brought the poorhouses to empty out. One by one through the 1950s, the poorhouses closed, responsibility for those who'd been classified as elderly "paupers" was transferred to departments of welfare, and the sick and disabled were put in hospitals. But hospitals couldn't solve the debilities of chronic illness and advancing age, and they began to fill up with people who had nowhere to go. The hospitals lobbied the government for help, and in 1954 lawmakers provided funding to enable them to build separate custodial units for patients needing an extended period of "recovery." That was the beginning of the modern nursing home. They were never created to help people facing dependency in old age. They were created to clear out hospital beds—which is why they were called "nursing" homes.

This has been the persistent pattern of how modern society has dealt with old age. The systems we've devised were almost always designed to solve some other problem. As one scholar put it, describing the history of nursing homes from the perspective of the elderly "is like describing the opening of the American West from the perspective of the mules; they were certainly there, and the epochal events were certainly critical to the mules, but hardly anyone was paying very much attention to them at the time."

The next major spur to American nursing home growth was similarly unintentional. When Medicare, America's health insurance system for the aged and disabled, passed in 1965, the law specified that it would pay only for care in facilities that met basic health and safety standards. A significant number of hospitals, especially in the South, couldn't meet those standards. Policy makers feared a major backlash from elderly patients with Medicare cards being turned away from their local hospital. So the Bureau of Health Insurance invented the concept of "substantial compliance"—if the hospital came "close" to meeting the standards and aimed to improve, it would be approved. The category was a complete fabrication with no legal basis, though it solved a problem without major harm—virtually all of the hospitals did improve. But the bureau's ruling gave an opening to nursing homes, few of which met even minimum federal standards such as having a nurse on-site or fire protections in place. Thousands of them, asserting that they were in "substantial compliance," were approved, and the number of nursing homes exploded—by 1970, some thirteen thousand of them had been built—and so did reports of neglect and mistreatment. That year in Marietta, Ohio, the next county over from my hometown, a nursing home fire trapped and killed thirty-two residents. In Baltimore, a *Salmonella* epidemic in a nursing home claimed thirty-six lives.

With time, regulations were tightened. The health and safety problems were finally addressed. Nursing homes are no longer firetraps. But the core problem persists. This place where half of us will typically spend a year or more of our lives was never truly made for us.

ONE MORNING IN late 1993, Alice had a fall while alone in her apartment. She wasn't found until many hours later when Nan, who was puzzled at not being able to reach her by phone, sent Jim to investigate. He discovered Alice laid out beside the living room couch, nearly unconscious. At the hospital, the medical team gave her intravenous fluids and a series of tests and X-rays. They found no broken bones or head injury. Everything seemed okay. But they also found no explanation for her fall beyond general frailty.

When she returned to Longwood House, she was encouraged to move to the skilled nursing floor. She resisted vehemently. She did not want to go. The staff relented. They checked her more frequently. Mary increased the hours she spent looking after her.

But before long, Jim got a call that Alice had fallen again. It was a bad fall, they said. She'd been taken by ambulance to a hospital. By the time he got there, she had already been wheeled into surgery. X-rays showed she'd broken her hip—the top of her femur had snapped like a glass stem. The orthopedic surgeons repaired the fracture with a couple of long metal nails.

This time, she came back to Longwood House in a wheelchair and needed help with virtually all of her everyday activities—using the toilet, bathing, dressing. Alice was left with no choice but to move into the skilled nursing unit. The hope, they told her, was that, with physical therapy, she'd learn to walk again and return to her apartment. But she never did. From then on, she was confined to a wheelchair and the rigidity of nursing home life.

All privacy and control were gone. She was put in hospital clothes most of the time. She woke when they told her, bathed and dressed when they told her, ate when they told her. She lived with whomever they said she had to. There was a succession of roommates, never chosen with her input and all with cognitive impairments. Some were quiet. One kept her up at night. She felt incarcerated, like she was in prison for being old.

The sociologist Erving Goffman noted the likeness between prisons and nursing homes half a century ago in his book *Asylums*. They were, along with military training camps, orphanages, and mental hospitals, "total institutions"—places largely cut off from wider society. "A basic social arrangement in modern society is that the individual tends to sleep, play, and work in different places, with different co-participants, under different authorities, and without an over-all rational plan," he wrote. By contrast, total institutions break down the barriers separating our spheres of life in specific ways that he enumerated:

First, all aspects of life are conducted in the same place and under the same central authority. Second, each phase of the member's daily activity is carried on in the immediate company of a large batch of others, all of whom are treated alike and required to do the same thing together. Third, all phases of the day's activities are tightly scheduled, with one activity leading at a prearranged time into the next, the whole sequence of activities being imposed from above by a system of explicit formal rulings and a body of officials. Finally, the various enforced activities are brought together into a single plan purportedly designed to fulfill the official aims of the institution.

In a nursing home, the official aim of the institution is caring, but the idea of caring that had evolved didn't bear any meaningful resemblance to what Alice would call living. She was hardly alone in feeling this way. I once met an eighty-nine-year-old woman who had, of her own volition, checked herself into a Boston nursing home. Usually, it's the children who push for a change, but in this case she was the one who did. She had congestive heart failure, disabling arthritis, and after a series of falls she felt she had little choice but to leave her condominium in Delray Beach, Florida. "I fell twice in one week, and I told my daughter I don't belong at home anymore," she said.

She picked the facility herself. It had excellent ratings and nice staff, and her daughter lived nearby. She had moved in the month before I met her. She told me she was glad to be in a safe place—if there's anything a decent nursing home is built for, it is safety. But she was wretchedly unhappy.

The trouble was that she expected more from life than safety.

"I know I can't do what I used to," she said, "but this feels like a hospital, not a home."

It is a near-universal reality. Nursing home priorities are matters like avoiding bedsores and maintaining residents' weight—important medical goals, to be sure, but they are means, not ends. The woman had left an airy apartment she furnished herself for a small beige hospital-like room with a stranger for a roommate. Her belongings were stripped down to what she could fit into the one cupboard and shelf they gave her. Basic matters, like when she went to bed, woke up, dressed, and ate, were subject to the rigid schedule of institutional life. She couldn't have her own furniture or a cocktail before dinner, because it wasn't safe.

There was so much more she felt she could do in her life. "I want to be helpful, play a role," she said. She used to make her own jewelry, volunteer at the library. Now, her main activities were bingo, DVD movies, and other forms of passive group entertainment. The things she missed most, she told me, were her friendships, privacy, and a purpose to her days. Nursing homes have come a long way from the firetrap warehouses of neglect they used to be. But it seems we've succumbed to a belief that, once you lose your physical independence, a life of worth and freedom is simply not possible.

The elderly themselves have not completely succumbed, however. Many resist. In every nursing home and assisted living facility, battles rage over the priorities and values people are supposed to live by. Some, like Alice, resist mainly through noncooperation—refusing the scheduled activities or medications. They are the ones we call "feisty." It's a favorite word for the aged. Outside a nursing home, we usually apply the adjective with a degree of admiration. We like the tenacious, sometimes cantankerous ways in which the Harry Trumans of the world assert themselves. But inside, when we say someone is feisty, we mean it in a less complimentary way. Nursing home staff like, and approve of, residents who are "fighters" and show "dignity and self-esteem"—until these traits interfere with the staff's priorities for them. Then they are "feisty."

Talk to the staff members and you will hear about the daily skirmishes. A woman calls for help to the bathroom "every five minutes." So they put her on a set schedule, taking her to the bathroom once every couple hours, when it fits into their rounds. But she doesn't go according to schedule, instead wetting her bed ten minutes after a bathroom trip. So now they put her in a diaper. Another resident refuses to use his walker and takes unauthorized, unaccompanied walks. A third sneaks cigarettes and alcohol.

Food is the Hundred Years' War. A woman with severe Parkinson's disease keeps violating her pureed diet restriction, stealing food from other residents that could cause her to choke. A man with Alzheimer's disease hoards snacks in his room, violating house rules. A diabetic is found eating clandestine sugar cookies and pudding, knocking his blood sugar levels off his target. Who knew you could rebel just by eating a cookie?

In the horrible places, the battle for control escalates until you get tied down or locked into your Geri-chair or chemically subdued with psychotropic medications. In the nice ones, a staff member cracks a joke, wags an affectionate finger, and takes your brownie stash away. In almost none does anyone sit down with you and try to figure out what living a life really means to you under the circumstances, let alone help you make a home where that life becomes possible.

This is the consequence of a society that faces the final phase of the human life cycle by trying not to think about it. We end up with institutions that address any number of societal goals—from freeing up hospital beds to taking burdens off families' hands to coping with poverty among the elderly—but never the goal that matters to the people who reside in them: how to make life worth living when we're weak and frail and can't fend for ourselves anymore.

ONE DAY WHEN Jim visited Alice, she whispered something in his ear. It was winter 1994, a few weeks after her hip fracture and admission to the skilled nursing unit and two years since she'd begun living at Longwood House. He'd wheeled her from her room for a stroll around the complex. They found a comfortable place in the lobby and stopped to sit for a while. They were both quiet people, and they'd been content to sit there silently, watching people come and go. That was when she leaned toward him in her wheelchair. She whispered just two words.

"I'm ready," she said.

He looked at her. She looked at him. And he understood. She was ready to die.

"Okay, Mom," Jim said.

It saddened him. He wasn't sure what to do about it. But not long afterward, the two of them arranged for a Do Not Resuscitate order to be put on record at the nursing home. If her heart or her breathing stopped, they would not attempt to rescue her from death. They would not do chest compressions or shock her or put a breathing tube down her throat. They would let her go.

Months passed. She waited and endured. One April night, she developed abdominal pains. She mentioned them briefly to a nurse, then decided to say nothing more. Later, she vomited blood. She alerted no one. She didn't press the call button or say anything to her roommate. She stayed in bed, silent. The next morning, when the aides came to wake the residents on her floor, they found she was gone.

4 · *Assistance*

You'd think people would have rebelled. You'd think we would have burned the nursing homes to the ground. We haven't, though, because we find it hard to believe that anything better is possible for when we are so weakened and frail that managing without help is no longer feasible. We haven't had the imagination for it.

In the main, the family has remained the primary alternative. Your chances of avoiding the nursing home are directly related to the number of children you have, and, according to what little research has been done, having at least one daughter seems to be crucial to the amount of help you will receive. But our greater longevity has coincided with the increased dependence of families on dual incomes, with results that are painful and unhappy for all involved.

Lou Sanders was eighty-eight years old when he and his daughter, Shelley, were confronted with a difficult decision about the future. Up to that point he had managed well. He'd never demanded much from life beyond a few modest pleasures and the company of family and friends. The son of Russian-speaking

Jewish immigrants from Ukraine, he'd grown up in Dorchester, a working-class neighborhood in Boston. In World War II, he served in the air force in the South Pacific, and after returning he married and settled in Lawrence, an industrial town outside Boston. He and his wife, Ruth, had a son and a daughter, and he went into the appliance business with a brother-in-law. Lou was able to buy the family a three-bedroom house in a nice neighborhood and give his children college educations. He and Ruth encountered their share of life's troubles. Their son, for instance, had serious problems with drugs, alcohol, and money and proved to have bipolar disorder. In his forties, he committed suicide. And the appliance business, which had done well for years, went belly-up when the chain stores came along. At fifty years old, Lou found himself having to start over. Nonetheless, despite his age, lack of experience, and lack of a college education, he was given a new chance as an electronic technician at Raytheon and ended up spending the remainder of his career there. He retired at sixty-seven, having worked the additional two years to get 3 percent extra on his Raytheon pension.

Meanwhile, Ruth developed health issues. A lifelong smoker, she was diagnosed with lung cancer, survived it, and kept smoking (which Lou couldn't understand). Three years after Lou retired, she had a stroke that she never wholly recovered from. She became increasingly dependent on him—for transportation, for shopping, for managing the house, for everything. Then she developed a lump under her arm, and a biopsy revealed metastatic cancer. She died in October 1994, at the age of seventy-three. Lou, at seventy-six, became a widower.

Shelley worried for him. She didn't know how he would get along without Ruth. Caring for Ruth through her decline, however, had forced him to learn to fend for himself, and, although he mourned, he gradually found that he didn't mind being on his own. For the next decade, he led a happy, satisfying life. He had a simple routine. He rose early in the morning, fixed himself breakfast, and read the newspaper. He'd take a walk, buy his groceries for the day at the supermarket, and come home to make his lunch. Later in the afternoon, he would go to the town library. It was pretty, light-filled, and quiet, and he'd spend a couple hours reading his favorite magazines and newspapers or burrowing into a thriller. Returning home, he'd read a book he'd checked out or watch a movie or listen to music. A couple of nights a week, he'd play cribbage with one of his neighbors in the building.

"My father developed really interesting friendships," Shelley said. "He could make friends with anyone."

One of Lou's new companions was an Iranian clerk at a video store in town where Lou often stopped in. The clerk, named Bob, was in his twenties. Lou would perch on a bar stool that Bob set up by the counter for him, and the two of them—the young Iranian and the old Jew—could hang out for hours. They became such good pals that they even traveled to Las Vegas together once. Lou loved going to casinos and made trips with an assortment of friends.

Then, in 2003, at the age of eighty-five, he had a heart attack. He proved lucky. An ambulance sped him to the hospital, and the doctors were able to stent open his blocked coronary artery in time. After a couple weeks in a cardiac rehabilitation center, it was as if nothing had happened at all. Three years later, however, he had his first fall—that harbinger of unstoppable trouble. Shelley noticed that he had developed a tremor, and a neurologist diagnosed him with Parkinson's disease. Medications controlled the symptoms, but he also began having trouble with his memory. Shelley observed that when he told a long story he sometimes lost the thread of what he was saying. Other times, he seemed confused about something they'd just spoken about. Most of the time he seemed fine, even exceptional for a man of eighty-eight years. He still drove. He still beat everyone at cribbage. He still looked after his home and managed his finances by himself. But then he had another bad fall, and it scared him. He suddenly felt the weight of all the changes that had been accumulating. He told Shelley he was afraid he might fall one day, hit his head, and die. It wasn't dying that scared him, he said, but the possibility of dying alone.

She asked him what he would think about looking at retirement homes. He wanted no part of it. He'd seen friends in those sorts of places.

"They're full of old people," he said. It was not the way he wanted to live. He made Shelley promise to never put him in such a place.

Still, he could no longer manage on his own. The only choice left for him was to move in with her and her family. So that's what Shelley arranged for him to do.

I asked her and her husband, Tom, how they had felt about this. Good, they both said. "I didn't feel comfortable with him living independently anymore," Shelley said, and Tom agreed. Lou'd had a heart attack. He was going on ninety. This was the least they could do for him. And, they admitted thinking, how long were they really going to have with him, anyway?

TOM AND SHELLEY lived comfortably in a modest colonial in North Reading, a Boston suburb, but never completely so. Shelley worked as a personal assistant. Tom had just spent a year and half unemployed after a layoff. Now he worked for a travel company for less than he used to earn. With two teenage children in the house, there was no obvious space for Lou. But Shelley and

Tom converted their living room into a bedroom, moving in a bed, an easy chair, Lou's armoire, and a flat-screen television. The rest of his furniture was sold off or put in storage.

Cohabitation required adjustment. Everyone soon discovered the reasons that generations prefer living apart. Parent and child traded roles, and Lou didn't like not being the master of his home. He also found himself lonelier than he expected. On their suburban cul-de-sac, he had no company for long stretches of the day and nowhere nearby to walk to—no library or video store or supermarket.

Shelley tried to get him involved in a day program for senior citizens. She took him to a breakfast they had. He didn't like it one bit. She discovered they made occasional trips to Foxwoods, a casino two hours from Boston. It wasn't his favorite, but he agreed to go. She was thrilled. She hoped he'd make friends.

She told me, "It felt like I was putting my child on the bus"—which was probably exactly what he disliked about it. "I remember saying, 'Hi, everyone. This is Lou. This is his first time so I hope you will all be friends with him.'" When he came back, she asked him if he'd made any friends. No, he said. He just gambled by himself.

Gradually, though, he found ways to adapt. Shelley and Tom had a Chinese Shar-Pei named Beijing, and Lou and the dog became devoted companions. She slept on his bed with him at night and sat with him when he read or watched TV. He took her on walks. If she was in his recliner, he'd go get another chair from the kitchen rather than disturb her.

He found human companions, too. He took to greeting the mailman each day, and they became friends. The mailman played cribbage, and he started coming over every Monday to play on his lunch hour. Shelley hired a young man named Dave to spend time with Lou, as well. It was the sort of preengineered playdate that is always doomed to failure, but—go figure—they hit it off. Lou played cribbage with Dave, too, and he came over a couple afternoons a week to hang out.

Lou settled in and imagined that this would be how he'd live out the rest of his days. But while he managed to adjust, Shelley found the situation steadily more impossible. She was working, looking after the home, and worrying about her kids, who had their own struggles as they made their way through high school. And then she had to look after her dear but frighteningly frail and dependent father. It was an enormous burden. The falls, for example, never stopped. He'd be in his room or in the bathroom or getting up from the kitchen table, when he'd suddenly pitch off his feet like a tree falling. In one year, he had four ambulance rides to the emergency room. The doctors stopped his Parkinson's medication, thinking that might be the culprit. But that only worsened his tremors and made him yet more unsteady on his feet. Eventually, he was diagnosed with postural hypotension—a condition of old age in which the body loses its ability to maintain adequate blood pressure for brain function during changes in position like standing up from sitting. The only thing the doctors could do was to tell Shelley to be more careful with him.

At night, she discovered, Lou had night terrors. He dreamt of war. He'd never been in hand-to-hand combat, but in his dreams an enemy would be attacking him with a sword, stabbing him or chopping his arm off. They were vivid and terrifying. He'd thrash and shout and hit the wall next to him. The family could hear him across the house: "Nooo!" "What do you mean?" "You son of a bitch!"

"We'd never heard him say anything like that before," Shelley said. He kept the family up many nights.

The demands on Shelley only mounted. At ninety, Lou no longer had the balance and dexterity required to bathe himself.

On the advice of a senior services program, Shelley installed bathroom grab bars, a sitting-height toilet, and a shower chair, but they weren't enough, so she arranged for a home health aide to help with washing and other tasks. But Lou didn't want showers in the daytime when an aide could help. He wanted baths in the nighttime, which required Shelley's help. So every day, this became her job, too.

It was the same with changing his clothes when he had wet himself. He had prostate issues, and, although the urologist gave him medicines for it, he still had problems with dribbles and leaks and not making it to the bathroom in time. Shelley tried to get him to wear protective disposable underwear, but he wouldn't do it. "They're diapers," he said.

The burdens were large and small. He didn't like the food she made for the rest of her family. He never complained. He just wouldn't eat. So she had to start making separate meals for him. He was hard of hearing and would blast the television in his room at brain-broiling volume. They'd shut his door, but he didn't like that—the dog couldn't get in and out. Shelley was ready to throttle him. Eventually, she found wireless earbuds called "TV ears." Lou hated them, but she made him use them. "They were a lifesaver," Shelley said. I wasn't sure if she meant that it was her life that they saved or his.

Taking care of a debilitated, elderly person in our medicalized era is an overwhelming combination of the technological and the custodial. Lou was on numerous medications, which had to be tracked and sorted and refilled. He had a small platoon of specialists he had to visit—at times, nearly weekly—and they were forever scheduling laboratory tests, imaging studies, and visits to other specialists. He had an electronic alert system for falls, which had to be tested monthly. And there was almost no help for Shelley. The burdens for today's caregiver have actually increased from what they would have been a century ago. Shelley had become a round-the-clock concierge/chauffeur/schedule manager/medication-and-technology troubleshooter, in addition to cook/maid/attendant, not to mention income earner. Last-minute cancellations by health aides and changes in medical appointments played havoc with her performance at work, and everything played havoc with her emotions at home. Just to take an overnight trip with her family, she had to hire someone to stay with Lou, and even then a crisis would scuttle the plans. One time, she went on a Caribbean vacation with her husband and kids but had to return after just three days. Lou needed her.

She felt her sanity slipping. She wanted to be a good daughter. She wanted her father to be safe, and she wanted him to be happy. But she wanted a manageable life, too. One night she asked her husband, should we find a place for him? She felt ashamed just voicing the thought. It'd break her promise to her father.

Tom wasn't much help. "You'll manage," he told her. "How much more time is there?"

Lots, it would turn out. "I was being insensitive to her," Tom told me, looking back three years later. Shelley was reaching the breaking point.

She had a cousin who ran an elder care organization. He recommended a nurse to come out to assess Lou and talk to him, so that Shelley didn't have to be the bad guy. The nurse told Lou that given his increased needs, he needed more help than he could get at home. He shouldn't be so alone through the day, she said.

He looked at Shelley imploringly, and she knew what he was thinking. Couldn't she just stop working and be there for him? The question felt like a dagger in her chest. Shelley teared up and told him that she couldn't provide the care he needed—

not emotionally and not financially. Reluctantly, he agreed to let her take him to look for a place. It seemed as if, once aging led to debility, it was impossible for anyone to be happy.

THE PLACE THEY decided to visit wasn't a nursing home but an assisted living facility. Today, assisted living is regarded as something of an intermediate station between independent living and life in a nursing home. But when Keren Brown Wilson, one of the originators of the concept, built her first assisted living home for the aged in Oregon in the 1980s, she was trying to create a place that would eliminate the need for nursing homes altogether. She'd wanted to build an alternative, not a halfway station. Wilson believed she could create a place where people like Lou Sanders could live with freedom and autonomy no matter how physically limited they became. She thought that just because you are old and frail, you shouldn't have to submit to life in an asylum. In her head she had a vision of how to make a better life achievable. And that vision had been formed by the same experiences—of reluctant dependency and agonized responsibility—that Lou and Shelley were grappling with.

The bookish daughter of a West Virginia coal miner and a washerwoman, neither of whom were schooled past eighth grade, Wilson was an unlikely radical. When she was in grade school, her father died. Then, when she was nineteen years old, her mother, Jessie, suffered a devastating stroke. Jessie was just fifty-five years old. The stroke left her permanently paralyzed down one side of her body. She could no longer walk or stand. She couldn't lift her arm. Her face sagged. Her speech slurred. Although her intelligence and perception were unaffected, she couldn't bathe herself, cook a meal, manage the toilet, or do her own laundry—let alone any kind of paid work. She needed help. But Wilson was just a college student. She had no income, a tiny apartment she shared with a roommate, and no way to take care of her mother. She had siblings but they were little better equipped. There was nowhere for Jessie but a nursing home. Wilson arranged for one near where she was in college. It seemed a safe and friendly place. But Jessie never stopped asking her daughter to "Take me home."

"Get me out of here," she said over and over again.

Wilson became interested in policy for the aged. When she graduated, she got a job working in senior services for the state of Washington. As the years passed, Jessie shifted through a series of nursing homes, near one or another of her children. She didn't like a single one of those places. Meanwhile, Wilson got married, and her husband, a sociologist, encouraged her to continue with her schooling. She was accepted as a PhD student in gerontology at Portland State University in Oregon. When she told her mother she would be studying the science of aging, Jessie asked her a question that Wilson says changed her life: "Why don't you do something to help people like me?"

"Her vision was simple," Wilson wrote later.

She wanted a small place with a little kitchen and a bathroom. It would have her favorite things in it, including her cat, her unfinished projects, her Vicks VapoRub, a coffeepot, and cigarettes. There would be people to help her with the things she couldn't do without help. In the imaginary place, she would be able to lock her door, control her heat, and have her own furniture. No one would make her get up, turn off her favorite soaps, or ruin her clothes. Nor could anyone throw out her "collection" of back issues and magazines and Goodwill treasures because they were a safety hazard. She could have privacy when-ever she wanted, and no one could make her get dressed, take her medicine, or go to activities she did not like. She would be Jessie again, a person living in an apartment instead of a patient in a bed.

Wilson didn't know what to do when her mother told her these things. Her mother's desires seemed both reasonable and—according to the rules of the places she'd lived—impossible. Wilson felt badly for the nursing home staff, who worked hard taking care of her mother and were just doing what they were expected to do, and she felt guilty that she couldn't do more herself. In graduate school, her mother's uncomfortable question nagged at her. The more she studied and probed, the more con-vinced she became that nursing homes would not accept any-thing like what Jessie envisioned. The institutions were designed in every detail for the control of their residents. The fact that this design was supposed to be for their health and safety—for their benefit—made the places only that much more benighted and impervious to change. Wilson decided to try spelling out on paper an alternative that would let frail elderly people maintain as much control over their care as possible, instead of having to let their care control them.

The key word in her mind was *home*. Home is the one place where your own priorities hold sway. At home, *you* decide how you spend your time, how you share your space, and how you manage your possessions. Away from home, you don't. This loss of freedom was what people like Lou Sanders and Wilson's mother, Jessie, dreaded.

Wilson and her husband sat at their dining table and began sketching out the features of a new kind of home for the elderly, a place like the one her mother had pined for. Then they tried to get someone to build it and test whether it would work. They approached retirement communities and builders. None were interested. The ideas seemed impractical and absurd. So the couple decided to build the place on their own.

They were two academics who had never attempted anything of the sort. But they learned one step at a time. They worked with an architect to lay out the plans in detail. They went to bank after bank to get a loan. When that did not succeed, they found a private investor who backed them but required them to give up majority ownership and to accept personal liability for failure. They signed the deal. Then the state of Oregon threatened to withhold licensing as senior housing because the plans stipulated that people with disabilities would be living there. Wilson spent several days camped out in one government office after another until she had secured an exemption. Unbelievably, she and her husband cleared every obstacle. And in 1983, their new "living center with assistance" for the elderly—named Park Place—opened in Portland.

By the time it opened, Park Place had become far more than an academic pilot project. It was a major real estate development with 112 units, and they filled up almost immediately. The concept was as appealing as it was radical. Although some of the residents had profound disabilities, none were called patients. They were all simply tenants and were treated as such. They had private apartments with a full bath, kitchen, and a front door that locked (a touch many found particularly hard to imagine). They were allowed to have pets and to choose their own carpeting and furniture. They were given control over temperature settings, food, who came into their home and when. They were just people living in an apartment, Wilson insisted over and over again. But, as elders with advancing disabilities, they were also provided with the sorts of help that my grandfather found so readily with his family all around. There was help with the basics—food, personal care, medications. There was a nurse on-site and tenants had a button for summoning urgent assistance at any time of day or night. There was also help with maintaining a decent quality of life—having company, keeping up their connections in the outside world, continuing the activities they valued most.

The services were, in most ways, identical to the services that nursing homes provide. But here the care providers understood they were entering someone else's home, and that changed the power relations fundamentally. The residents had control over the schedule, the ground rules, the risks they did and didn't want to take. If they wanted to stay up all night and sleep all day, if they wanted to have a gentleman or lady friend stay over, if they wanted not to take certain medications that made them feel groggy; if they wanted to eat pizza and M&M's despite swallowing problems and no teeth and a doctor who'd said they should eat only pureed glop—well, they could. And if their mind had faded to the point that they could no longer make rational decisions, then their family—or whomever they'd designated—could help negotiate the terms of the risks and choices that were acceptable. With "assisted living," as Wilson's concept become known, the goal was that no one ever had to feel institutionalized.

The concept was attacked immediately. Many longtime advocates for the protection of the elderly saw the design as fundamentally dangerous. How was the staff going to keep people safe behind closed doors? How could people with physical disabilities and memory problems be permitted to have cooktops, cutting knives, alcohol, and the like? Who was going to ensure that the pets they chose were safe? How was the carpeting going to be sanitized and kept free of urine odors and bacteria?

How would the staff know if a tenant's health condition had changed?

These were legitimate questions. Is someone who refuses regular housekeeping, smokes cigarettes, and eats candies that cause a diabetic crisis requiring a trip to the hospital someone who is a victim of neglect or an archetype of freedom? There is no clean dividing line, and Wilson was not offering simple answers. She held herself and her staff responsible for developing ways of ensuring the safety of the tenants. At the same time, her philosophy was to provide a place where residents retained the autonomy and privacy of people living in their own homes—including the right to refuse strictures imposed for reasons of safety or institutional convenience.

The state monitored the experiment closely. When the group expanded to a second location in Portland—this one had 142 units and capacity for impoverished elderly people on government support—the state required Wilson and her husband to track the health, cognitive capabilities, physical function, and life satisfaction of the tenants. In 1988, the findings were made public. They revealed that the residents had not in fact traded their health for freedom. Their satisfaction with their lives increased, and at the same time their health was maintained. Their physical and cognitive functioning actually improved. Incidence of major depression fell. And the cost for those on government support was 20 percent lower than it would have been in a nursing home. The program proved an unmitigated success.

AT THE CENTER of Wilson's work was an attempt to solve a deceptively simple puzzle: what makes life worth living when we are old and frail and unable to care for ourselves? In 1943, the psychologist Abraham Maslow published his hugely influential paper "A Theory of Human Motivation," which famously described people as having a hierarchy of needs. It is often depicted as a pyramid. At the bottom are our basic needs—the essentials of physiological survival (such as food, water, and air) and of safety (such as law, order, and stability). Up one level are the need for love and for belonging. Above that is our desire for growth—the opportunity to attain personal goals, to master knowledge and skills, and to be recognized and rewarded for our achievements. Finally, at the top is the desire for what Maslow termed "self-actualization"—self-fulfillment through pursuit of moral ideals and creativity for their own sake.

Maslow argued that safety and survival remain our primary and foundational goals in life, not least when our options and capacities become limited. If true, the fact that public policy and concern about old age homes focus on health and safety is just a recognition and manifestation of those goals. They are assumed to be everyone's first priorities.

Reality is more complex, though. People readily demonstrate a willingness to sacrifice their safety and survival for the sake of something beyond themselves, such as family, country, or justice. And this is regardless of age.

What's more, our driving motivations in life, instead of remaining constant, change hugely over time and in ways that don't quite fit Maslow's classic hierarchy. In young adulthood, people seek a life of growth and self-fulfillment, just as Maslow suggested. Growing up involves opening outward. We search out new experiences, wider social connections, and ways of putting our stamp on the world. When people reach the latter half of adulthood, however, their priorities change markedly. Most reduce the amount of time and effort they spend pursuing achievement and social networks. They narrow in. Given the choice, young people prefer meeting new people to spending time with, say, a sibling; old people prefer the opposite. Studies find that as people grow older they interact with fewer people and concentrate more on spending time with family and established friends. They focus on being rather than doing and on the present more than the future.

Understanding this shift is essential to understanding old age. A variety of theories have attempted to explain why the shift occurs. Some have argued that it reflects wisdom gained from long experience in life. Others suggest it is the cognitive result of changes in the tissue of the aging brain. Still others argue that the behavior change is forced upon the elderly and does not actually reflect what they want in their heart of hearts. They narrow in because the constrictions of physical and cognitive decline prevent them from pursuing the goals they once had or because the world stops them for no other reason than they are old. Rather than fight it, they adapt—or, to put it more sadly, they give in.

Few researchers in recent decades have done more creative or important work sorting these arguments out than the Stanford psychologist Laura Carstensen. In one of her most influential studies, she and her team tracked the emotional experiences of nearly two hundred people over years of their lives. The subjects spanned a broad range of backgrounds and ages. (They were from eighteen to ninety-four years old when they entered the study.) At the beginning of the study and then every five years, the subjects were given a beeper to carry around twenty-four hours a day for one week. They were randomly paged thirty-five times over the course of that week and asked to choose from a list all the emotions they were experiencing at that exact moment.

If Maslow's hierarchy was right, then the narrowing of life

runs against people's greatest sources of fulfillment and you would expect people to grow unhappier as they age. But Carstensen's research found exactly the opposite. The results were unequivocal. Far from growing unhappier, people reported more positive emotions as they aged. They became less prone to anxiety, depression, and anger. They experienced trials, to be sure, and more moments of poignancy—that is, of positive and negative emotion mixed together. But overall, they found living to be a more emotionally satisfying and stable experience as time passed, even as old age narrowed the lives they led.

The findings raised a further question. If we shift as we age toward appreciating everyday pleasures and relationships rather than toward achieving, having, and getting, and if we find this more fulfilling, then why do we take so long to do it? Why do we wait until we're old? The common view was that these lessons are hard to learn. Living is a kind of skill. The calm and wisdom of old age are achieved over time.

Carstensen was attracted to a different explanation. What if the change in needs and desires has nothing to do with age per se? Suppose it merely has to do with perspective—your personal sense of how finite your time in this world is. This idea was regarded in scientific circles as somewhat odd. But Carstensen had her own reason for thinking that one's personal perspective might be centrally important—a near-death experience that radically changed her viewpoint on her own life.

It was 1974. She was twenty-one, with an infant at home and a marriage already in divorce proceedings. She had only a high school education and a life that no one—least of all she— would have predicted might someday lead to an eminent scientific career. But one night, she left the baby with her parents and went out with friends to party and see the band Hot Tuna in concert. Coming back from the show, they piled into a VW minibus, and, on a highway somewhere outside Rochester, New York, the driver, drunk, rolled the minibus over an embankment.

Carstensen barely survived. She had a serious head injury, internal bleeding, multiple shattered bones. She spent months in the hospital. "It was that cartoonish scene, lying on my back, leg tied in the air," she told me. "I had a lot of time to think after the initial three weeks or so, when things were really touch and go and I was coming in and out of consciousness.

"I got better enough to realize how close I had come to losing my life, and I saw very differently what mattered to me. What mattered were other people in my life. I was twenty-one. Every thought I'd had before that was: What was I going to do next in life? And how would I become successful or not successful? Would I find the perfect soul mate? Lots of questions like that, which I think are typical of twenty-one-year-olds.

"All of a sudden, it was like I was stopped dead in the tracks. When I looked at what seemed important to me, very different things mattered."

She didn't instantly recognize how parallel her new perspective was to the one old people commonly have. But the four other patients in her ward were all elderly women—their legs strung up in the air after hip fractures—and Carstensen found herself connecting with them.

"I was lying there, surrounded by old people," she said. "I got to know them, see what was happening to them." She noticed how differently they were treated from her. "I basically had doctors and therapists coming in and working with me all day long, and they would sort of wave at Sadie, the lady in the next bed, on the way out and say, 'Keep up the good work, hon!'" The message was: This young woman's life had possibilities. Theirs didn't.

"It was this experience that led me to study aging," Carstensen said. But she didn't know at the time that it would. "I was not on a trajectory to end up being a professor at Stanford by any means at that point in my life." Her father, however, realized how bored she was lying there and took the opportunity to enroll her in a course at a local college. He went to all the lectures, audiotaped them, and brought the cassettes to her. She ended up taking her first college course in a hospital, on a women's orthopedics ward.

What was that first class, by the way? Introduction to Psychology. Lying there on that ward, she found she was living through the phenomena she was studying. Right from the start, she could see what the experts were getting right and what they were getting wrong.

Fifteen years later, when she was a scholar, the experience led her to formulate a hypothesis: how we seek to spend our time may depend on how much time we perceive ourselves to have. When you are young and healthy, you believe you will live forever. You do not worry about losing any of your capabilities. People tell you "the world is your oyster," "the sky is the limit," and so on. And you are willing to delay gratification—to invest years, for example, in gaining skills and resources for a brighter future. You seek to plug into bigger streams of knowledge and information. You widen your networks of friends and connections, instead of hanging out with your mother. When horizons are measured in decades, which might as well be infinity to human beings, you most desire all that stuff at the top of Maslow's pyramid—achievement, creativity, and other attributes of "self-actualization." But as your horizons contract—when you see the future ahead of you as finite and uncertain—your focus shifts to the here and now, to everyday pleasures and the people closest to you.

Carstensen gave her hypothesis the impenetrable name "socioemotional selectivity theory." The simpler way to say it is that perspective matters. She produced a series of experiments to test the idea. In one, she and her team studied a group of adult men ages twenty-three to sixty-six. Some of the men were healthy. But some were terminally ill with HIV/AIDS. The subjects were given a deck of cards with descriptions of people they might know, ranging in emotional closeness from family members to the author of a book they'd read, and they were asked to sort the cards according to how they would feel about spending half an hour with them. In general, the younger the subjects were, the less they valued time with people they were emotionally close to and the more they valued time with people who were potential sources of information or new friendship. However, among the ill, the age differences disappeared. The preferences of a young person with AIDS were the same as those of an old person.

Carstensen tried to find holes in her theory. In another experiment, she and her team studied a group of healthy people ages eight to ninety-three. When they were asked how they would like to spend half an hour of time, the age differences in their preferences were again clear. But when asked simply to imagine they were about to move far away, the age differences again disappeared. The young chose as the old did. Next, the researchers asked them to imagine that a medical breakthrough had been made that would add twenty years to their life. Again, the age differences disappeared—but this time the old chose as the young did.

Cultural differences were not significant, either. The findings in a Hong Kong population were identical to an American one. Perspective was all that mattered. As it happened, a year after the team had completed its Hong Kong study, the news came out that political control of the country would be handed over to China. People developed tremendous anxiety about what would happen to them and their families under Chinese rule. The researchers recognized an opportunity and repeated the survey. Sure enough, they found that people had narrowed their social networks to the point that the differences in the goals of young and old vanished. A year after the handover, when the uncertainty had subsided, the team did the survey again. The age differences reappeared. They did the study yet again after the 9/11 attacks in the United States and during the SARS epidemic that spread through Hong Kong in spring 2003, killing three hundred people in a matter of weeks. In each case the results were consistent. When, as the researchers put it, "life's fragility is primed," people's goals and motives in their everyday lives shift completely. It's perspective, not age, that matters most.

Tolstoy recognized this. As Ivan Ilyich's health fades and he realizes that his time is limited, his ambition and vanity disappear. He simply wants comfort and companionship. But almost no one understands—not his family, his friends, or the stream of eminent physicians whom his wife pays to examine him.

Tolstoy saw the chasm of perspective between those who have to contend with life's fragility and those who don't. He grasped the particular anguish of having to bear such knowledge alone. But he saw something else, as well: even when a sense of mortality reorders our desires, these desires are not impossible to satisfy. Although none of Ivan Ilyich's family or friends or doctors grasp his needs, his servant Gerasim does. Gerasim sees that Ivan Ilyich is a suffering, frightened, and lonely man and takes pity on him, aware that someday he himself would share his master's fate. While others avoid Ivan Ilyich, Gerasim talks to him. When Ivan Ilyich finds that the only position that relieves his pain is with his emaciated legs resting on Gerasim's shoulders, Gerasim sits there the entire night to provide comfort. He doesn't mind his role, not even when he has to lift Ilyich to and from the commode and clean up after him. He provides care without calculation or deception, and he doesn't impose any goals beyond what Ivan Ilyich desires. This makes all the difference in Ivan Ilyich's waning life:

> Gerasim did it all easily, willingly, simply, and with a good nature that touched Ivan Ilyich. Health, strength, and vitality in other people were offensive to him, but Gerasim's strength and vitality did not mortify but soothed him.

This simple but profound service—to grasp a fading man's need for everyday comforts, for companionship, for help achieving his modest aims—is the thing that is still so devastatingly lacking more than a century later. It was what Alice Hobson needed but could not find. And it was what Lou Sanders's daughter, through four increasingly exhausting years, discovered she could no longer give all by herself. But with the concept of assisted living, Keren Brown Wilson had managed to embed that vital help in a home.

THE IDEA SPREAD astoundingly quickly. Around 1990, based on Wilson's successes, Oregon launched an initiative to encourage the building of more homes like hers. Wilson worked with her husband to replicate their model and to help others do the same. They found a ready market. People proved willing to pay considerable sums to avoid ending up in a nursing home, and several states agreed to cover the costs for poor elders.

Not long after that, Wilson went to Wall Street for capital to build more places. Her company, Assisted Living Concepts, went public. Others sprang up with names like Sunrise, Atria, Sterling, and Karrington, and assisted living became the fastest-growing form of senior housing in the country. By 2000, Wilson had expanded her company from fewer than a hundred employees to more than three thousand. It operated 184 residences in eighteen states. By 2010, the number of people in assisted living was approaching the number in nursing homes.

But a distressing thing happened along the way. The concept of assisted living became so popular that developers began slapping the name on just about anything. The idea mutated from a radical alternative to nursing homes into a menagerie of watered-down versions with fewer services. Wilson testified before Congress and spoke across the country about her increasing alarm at the way the idea was evolving.

"With a general desire to adopt the name, suddenly assisted living was a redecorated wing of a nursing facility, or a sixteen-bed boarding house looking to attract private-pay clients," she reported. However much she attempted to uphold her founding philosophy, few others were as committed.

Assisted living most often became a mere layover on the way from independent living to a nursing home. It became part of the now widespread idea of a "continuum of care," which sounds perfectly nice and logical but manages to perpetuate conditions that treat the elderly like preschool children. Concern about safety and lawsuits increasingly limited what people could have in their assisted living apartments, mandated what activities they were expected to participate in, and defined ever more stringent move-out conditions that would trigger "discharge" to a nursing facility. The language of medicine, with its priorities of safety and survival, was taking over, again. Wilson pointed out angrily that even children are permitted to take more risks than the elderly. They at least get to have swings and jungle gyms.

A survey of fifteen hundred assisting living facilities published in 2003 found that only 11 percent offered both privacy and sufficient services to allow frail people to remain in residence. The idea of assisted living as an alternative to nursing homes had all but died. Even the board of Wilson's own company—having noted how many other companies were taking a less difficult and less costly direction—began questioning her standards and philosophy. She wanted to build smaller buildings, in smaller towns where elderly people had no options except nursing homes, and she wanted units for low-income elderly on Medicaid. But the more profitable direction was bigger buildings, in bigger cities, without low-income clientele or advanced services. She'd created assisted living to help people like her mother, Jessie, live a better life, and she'd shown that it could be profitable. But her board and Wall Street wanted avenues to even bigger profits. Her battles escalated until, in 2000, she stepped down as CEO and sold all her shares in the company she'd founded.

More than a decade has passed since. Keren Wilson has crossed into middle age. When I spoke to her not long ago, her crooked-toothed smile, slumped shoulders, reading glasses, and white hair made her look more like a bookish grandmother than the revolutionary entrepreneur who'd founded a worldwide industry. Ever the gerontologist, she gets excited when the conversation veers to research questions, and she is precise when she speaks. She nonetheless remains the sort of person who is perpetually in the grip of big, seemingly impossible problems. The company made her and her husband wealthy, and with their money they started the Jessie F. Richardson Foundation, named after her mother, in order to continue the work of transforming care for the elderly.

Wilson spends much of her time back in the West Virginia coal counties around where she was born—places like Boone and Mingo and McDowell. West Virginia has one of the oldest and poorest populations of any state in the country. As in so much of the world, it is a place where the young leave to seek better opportunities and the elderly are left behind. There, in the hollows where she grew up, Wilson is still trying to work out how ordinary people can age without having to choose between neglect and institutionalization. It remains among the most uncomfortable questions we face.

"I want you to know that I still love assisted living," she said, and she repeated herself: "I *love* assisted living." It had created a belief and expectation that there could be something better than a nursing home, she said, and it still does. Nothing that takes off becomes quite what the creator wants it to be. Like a child, it grows, not always in the expected direction. But Wilson continues to find places where her original intention remains alive.

"I love it when assisted living works," she said.

It's just that in most places it doesn't.

FOR LOU SANDERS, it didn't. Shelley felt lucky to find an assisted living facility near her home that would accept him with his meager finances. His savings were almost gone, and most other places expected upfront payments of hundreds of thousands of dollars. The home she found for Lou received government subsidies that made it affordable. It had a lovely porch, fresh paint, plenty of light in the lobby, a pretty library, and reasonably spacious apartments. It seemed inviting and professional. Shelley liked it from the first visit. But Lou resisted. He looked around and saw not a single person without a walker.

"I'll be the only one on my own two feet," he said. "It's not for me." They went back home.

Not long after, however, he had yet another fall. He went down hard in a parking lot, and his head took a sickening bounce on the asphalt. He didn't come to for a while. He was admitted to the hospital for observation. After that, he accepted that things had changed. He let Shelley put him on the waiting list for the assisted living facility. An opening came up just before his ninety-second birthday. If he didn't take the spot, they told him, he'd go to the bottom of the list. His hand was forced.

After the move, he wasn't angry with Shelley. But she might have found anger easier to deal with. He was just depressed, and what is a child to do about that?

Some of the problem, Shelley felt, was just the difficulty of dealing with change. At his age, Lou didn't do well with change. But she sensed that there was more to it than that. Lou looked lost. He didn't know a soul, and there was hardly another male to be found. He would look around thinking, What is a guy like me doing stuck in a place like this—with its bead-making workshops, cupcake-decorating afternoons, and crummy, Danielle Steel–filled library? Where was his family, or his friend the mailman, or Beijing, his beloved dog? He didn't belong. Shelley asked the activities director if she would plan a few activities that were more gender appropriate, maybe start a book club. But bah, like that was going to help.

What bothered Shelley most was how little curiosity the staff members seemed to have about what Lou cared about in his life and what he had been forced to forfeit. They didn't even recognize their ignorance in this regard. They might have called the service they provided assisted living, but no one seemed to think it was their job to actually assist him with living—to figure out how to sustain the connections and joys that most mattered to him. Their attitude seemed to result from incomprehension rather than cruelty, but, as Tolstoy would have said, what's the difference in the end?

Lou and Shelley worked out a compromise. She would bring him home every Sunday through Tuesday. That let him have something to look forward to each week and helped her feel better, too. At least, he'd have a couple days a week of the life he'd enjoyed.

I asked Wilson why assisted living so often fell short. She saw several reasons. First, to genuinely help people with living "is harder to do than to talk about" and it's difficult to make caregivers think about what it really entails. She gave the example of helping a person dress. Ideally, you let people do what they can themselves, thus maintaining their capabilities and sense of independence. But, she said, "Dressing somebody is easier than letting them dress themselves. It takes less time. It's less aggravation." So unless supporting people's capabilities is made a priority, the staff ends up dressing people like they're rag dolls. Gradually, that's how everything begins to go. The tasks come to matter more than the people.

Compounding matters, we have no good metrics for a place's success in assisting people to live. By contrast, we have very precise ratings for health and safety. So you can guess what gets the attention from the people who run places for the elderly: whether Dad loses weight, skips his medications, or has a fall, not whether he's lonely.

Most frustrating and important, Wilson said, assisted living isn't really built for the sake of older people so much as for the sake of their children. The children usually make the decision about where the elderly live, and you can see it in the way that places sell themselves. They try to create what the marketers call "the visuals"—the beautiful, hotel-like entryway, for instance, that caught Shelley's eye. They tout their computer lab, their exercise center, and their trips to concerts and museums—features that speak much more to what a middle-aged person desires for a parent than to what the parent does. Above all, they sell themselves as safe places. They almost never sell themselves as places that put a person's choices about how he or she wants to live first and foremost. Because it's often precisely the parents' cantankerousness and obstinacy about the choices they make that drive children to bring them on the tour to begin with. Assisted living has become no different in this respect than nursing homes.

A colleague once told her, Wilson said, "We want autonomy for ourselves and safety for those we love." That remains the main problem and paradox for the frail. "Many of the things that we want for those we care about are things that we would adamantly oppose for ourselves because they would infringe upon our sense of self."

She puts some of the blame on the elderly. "Older people are in part responsible for this because they disperse the decision making to their children. Part of it is an assumption about age and frailty, and it's also a bonding thing that goes on from older people to children. It's sort of like, 'Well, you're in charge now.'"

But, she said, "It's the rare child who is able to think, 'Is this place what Mom would want or like or need?' It's more like they're seeing it through their own lens." The child asks, "Is this a place *I* would be comfortable leaving Mom?"

Lou had not been in the assisted living home a year before

it became inadequate for him. He'd initially made the best of it. He discovered the one other Jewish guy in the place, a man named George, and they hit it off. They played cribbage and each Saturday went to temple, a routine Lou had endeavored all his life to avoid. Several of the ladies took special interest in him, which he mostly deflected. But not always. He had a little party one evening in his apartment, at which he was joined by two of his admirers and broke out a bottle of brandy he'd been given.

"Then my father passed out and hit his head on the floor and ended up in the emergency room," Shelley said. He could laugh about it later, when he got out of rehab. "Look at that," she recalled him saying. "I have the women over. Then one little drink, and I pass out."

Between the three days in Shelley's home each week and the pieces of a life Lou put together the rest of the week—the assisted living home's fecklessness notwithstanding—he was managing. Doing so had taken months. At ninety-two, he gradually rebuilt an everyday life he could abide.

His body wouldn't cooperate, though. His postural hypotension worsened. He passed out more frequently—not just when he had a brandy. It could be day or night, walking around or getting out of bed. There were multiple ambulance rides and trips to the doctor for X-rays. Things got to the point where he couldn't manage the long hallway and elevator to the dining room for meals anymore. He continued to refuse a walker. It was a point of pride. Shelley had to stock his refrigerator with prepared foods he could microwave.

She found herself worrying about him all over again. He wasn't eating properly. His memory was getting worse. And even with the regular health aide visits and evening checks, he was mostly sitting in his room by himself. She felt he didn't have enough supervision for how frail he was becoming. She would have to move him to somewhere with twenty-four-hour care.

She visited a nursing home nearby. "It was actually one of the nicer ones," she said. "It was clean." But it was a nursing home. "You had the people in their wheelchairs all slumped over and lined up in the corridors. It was horrible." It was the sort of place, she said, that her father feared more than anything. "He did not want his life reduced to a bed, a dresser, a tiny TV, and half of a room with the curtain between him and someone else."

But, she said, as she walked out of the place she thought, "This is what I have to do." Awful as it seemed, it was where she had to put him.

Why, I asked?

"To me, safety was paramount. That came before anything. I had to think of his safety," she said. Keren Wilson was right about the way the process evolves. Out of love and devotion, Shelley felt she had no choice but to put him where he dreaded.

I pressed her. Why? He had adjusted to where he was. He'd reassembled the pieces of a life—a friend, a routine, some things he still liked to do. It was true that he wasn't as safe as he would be in a nursing home. He still feared having that big fall and no one finding him before it was too late. But he was happier. And given his druthers, he'd choose the happier place. So why choose differently?

She didn't know how to answer. She found it hard to fathom any other way. He needed someone to look after him. He wasn't safe. Was she really supposed to just leave him there?

So this is the way it unfolds. In the absence of what people like my grandfather could count on—a vast extended family constantly on hand to let him make his own choices—our elderly are left with a controlled and supervised institutional existence, a medically designed answer to unfixable problems, a life designed to be safe but empty of anything they care about.

5 · *A Better Life*

In 1991, in the tiny town of New Berlin, in upstate New York, a young physician named Bill Thomas performed an experiment. He didn't really know what he was doing. He was thirty-one years old, less than two years out of family medicine residency, and he had just taken a new job as medical director of Chase Memorial Nursing Home, a facility with eighty severely disabled elderly residents. About half of them were physically disabled; four out of five had Alzheimer's disease or other forms of cognitive disability.

Up until then Thomas had worked as an emergency physician at a nearby hospital, the near opposite of a nursing home. People arrived in the emergency room with discrete, reparable problems—a broken leg, say, or a cranberry up the nose. If a patient had larger, underlying issues—if, for instance, the broken leg had been caused by dementia—his job was to ignore the issues or send the person somewhere else to deal with them, such as a nursing home. He took this new medical director job as a chance to do something different.

The staff at Chase saw nothing especially problematic about

the place, but Thomas with his newcomer's eyes saw despair in every room. The nursing home depressed him. He wanted to fix it. At first, he tried to fix it the way that, as a doctor, he knew best. Seeing the residents so devoid of spirit and energy, he suspected that some unrecognized condition or improper combination of medicines might be afflicting them. So he set about doing physical examinations of the residents and ordering scans and tests and changing their medications. But, after several weeks of investigations and alterations, he'd accomplished little except driving the medical bills up and making the nursing staff crazy. The nursing director talked to him and told him to back off.

"I was confusing care with treatment," he told me.

He didn't give up, though. He came to think the missing ingredient in this nursing home was life itself, and he decided to try an experiment to inject some. The idea he came up with was as mad and naïve as it was brilliant. That he got the residents and nursing home staff to go along with it was a minor miracle.

But to understand the idea—including how it came about and how he got it off the ground—you have to understand a few things about Bill Thomas. The first thing is that, as a child, Thomas won every sales contest his school had. They'd send the kids off to sell candles or magazines or chocolates door-to-door for the Boy Scouts or a sports team, and he'd invariably come home with the prize for most sales. He also won election as student body president in high school. He was chosen captain of the track team. When he wanted to, he could sell people on almost anything, including himself.

At the same time, he was a terrible student. He had miserable grades and repeated run-ins with his teachers over his failure to do the work they assigned. It wasn't that he couldn't do the work. He was a voracious reader and autodidact, the kind of a boy who would teach himself trigonometry so he could build a boat (which he did). He just didn't care about doing the work his teachers asked for, and he didn't hesitate to tell them so. Today, we'd diagnose him as having Oppositional Defiant Disorder. In the 1970s, they just thought he was trouble.

The two personas—the salesman and the defiant pain in the neck—seemed to come from the same place. I asked Thomas what his special technique for sales was as a kid. He said he didn't have any. It was simply that "I was willing to be rejected. That's what allows you to be a good salesperson. You have to be willing to be rejected." It was a trait that let him persist until he got what he wanted and avoid whatever he didn't want.

For a long time, though, he didn't know what he wanted. He had grown up in the next county over from New Berlin, in a valley outside the town of Nichols. His father had been a factory worker, his mother a telephone operator. Neither had gone to college, and no one expected Bill Thomas to go either. As he came to the end of high school, he was on track to join a union training program. But a chance conversation with a friend's older brother who was visiting home from college and told him about the beer, the girls, and the good times made him rethink.

He enrolled in a nearby state college, SUNY Cortland. There, something ignited him. Perhaps it was the high school teacher who predicted as he left that he'd be back in town pumping gas before Christmas. Whatever it was, he succeeded far beyond anyone's expectation, chewing through the curriculum, holding on to a 4.0 grade point average, and becoming student body president again. He had gone in thinking he might become a gym teacher, but in biology class he began thinking that maybe medicine was for him. He ended up becoming Cortland's first student to get into Harvard Medical School.

He loved Harvard. He could have gone there with a chip on his shoulder—the working-class kid out to prove he was nothing like those snobs, with their Ivy League educations and trust fund accounts. But he didn't. He found the place to be a revelation. He loved being with people who were so driven and passionate about science, medicine, everything.

"One of my favorite parts of medical school was that a group of us had dinner at the Beth Israel Hospital cafeteria every night," he told me. "And it would be two and a half hours of arguing cases—intense and really great."

He also loved being in a place where people believed he was capable of momentous things. Nobel Prize winners came to teach classes, even on Saturday mornings, because they expected him and the others to aspire to greatness.

He never felt the need to win anyone's approval, however. Faculty tried to recruit him to their specialized training programs at big-name hospitals or to their research laboratories. Instead, he chose family medicine residency in Rochester, New York. It wasn't exactly Harvard's idea of aspiring to greatness.

Returning home to upstate New York had been his goal all along. "I'm a local guy," he told me. In fact, his four years at Harvard were the only time he ever lived outside upstate New York. During vacations, he used to bicycle from Boston to Nichols and back—a 330-mile ride in each direction. He liked the self-sufficiency—pitching his tent in random orchards and fields along the road and finding food wherever he could. Family medicine was attractive in the same way. He could be independent, go it alone.

Partway through residency, when he'd saved up some money, he bought some farmland near New Berlin that he'd often passed on his bike rides and imagined owning some day. By the time he finished his training, working the land had become his real love. He entered local practice but soon focused on emergency medicine because it offered predictable hours, on a shift, letting him devote the rest of his time to his farm. He was committed to the idea of homesteading—being totally self-reliant. He built his home by hand with friends. He grew most of his own food. He used wind and solar power to generate electricity. He was completely off the grid. He lived by the weather and the seasons. Eventually, he and Jude, a nurse who became his wife, expanded the farm to more than four hundred acres. They had cattle, draft horses, chickens, a root cellar, a sawmill, and a sugarhouse, not to mention five children.

"I really felt that the life I was living was the most authentically true life I could live," Thomas explained.

He was at that point more farmer than doctor. He had a Paul Bunyan beard and was more apt to wear overalls beneath his white coat than a tie. But the emergency room hours were draining. "Basically, I got sick of working all those nights," he said. So he took the job in the nursing home. It was a day job. The hours were predictable. How hard could it be?

FROM THE FIRST day on the job, he felt the stark contrast between the giddy, thriving abundance of life that he experienced on his farm and the confined, institutionalized absence of life that he encountered every time he went to work. What he saw gnawed at him. The nurses said he would get used to it, but he couldn't, and he didn't want to go along with what he saw. Some years would pass before he could fully articulate why, but in his bones he recognized that the conditions at Chase Memorial Nursing Home fundamentally contradicted his ideal of self-sufficiency.

Thomas believed that a good life was one of maximum independence. But that was precisely what the people in the home were denied. He got to know the nursing home residents. They had been teachers, shopkeepers, housewives, and factory workers, just like people he'd known growing up. He was sure something better must be possible for them. So, acting on little more than instinct, he decided to try to put some life into the nursing home the way that he had done in his own home—by literally putting life into it. If he could introduce plants, animals, and children into the lives of the residents—fill the nursing home with them—what would happen?

He went to Chase's management. He proposed that they could fund his idea by applying for a small New York State grant that was available for innovations. Roger Halbert, the administrator who'd hired Thomas, liked the idea in principle. He was happy to try something new. During twenty years at Chase, he had ensured that the facility had an excellent reputation, and it had steadily expanded the range of activities available to the residents. Thomas's new idea seemed in line with past improvements. So the leadership team sat down together to write the application for the innovation funding. Thomas, however, seemed to have something in mind that was more extensive than Halbert had quite fathomed.

Thomas laid out the thinking behind his proposal. The aim, he said, was to attack what he termed the Three Plagues of nursing home existence: boredom, loneliness, and helplessness. To attack the Three Plagues they needed to bring in some life. They'd put green plants in every room. They'd tear up the lawn and create a vegetable and flower garden. And they'd bring in animals.

So far this sounded okay. An animal could sometimes be tricky because of health and safety issues. But nursing home regulations in New York permitted one dog or one cat. Halbert told Thomas that they'd tried a dog two or three times in the past without success. The animals had the wrong personality, and there were difficulties arranging for proper care. But he was willing to try again.

So Thomas said, "Let's try two dogs."

Halbert said, "The code doesn't allow that."

Thomas said, "Let's just put it down on paper."

There was silence for a moment. Even this small step pushed up against the values at the heart not just of nursing home regulations but also of what nursing homes believed they principally exist for—the health and safety of elders. Halbert had a hard time wrapping his mind around the idea. When I spoke to him not long ago, he still recalled the scene vividly.

> The director of nursing, Lois Greising, was sitting in the room, the activities leader, and the social worker. . . . And I can see the three of them sitting there, looking at each other, rolling their eyes, saying, "This is going to be interesting."
>
> I said, "All right, I'll put it down." I was beginning to think, "I'm not really into this as much as you are, but I'll put two dogs down."
>
> He said, "Now, what about cats?"
>
> I said, "What about cats?" I said, "We've got two dogs down on the paper."
>
> He said, "Some people aren't dog lovers. They like cats."
>
> I said, "You want dogs AND cats?"
>
> He said, "Let's put it down for discussion purposes."
>
> I said, "Okay. I'll put a cat down."
>
> "No, no, no. We're two floors. How about two cats on both floors?"
>
> I said, "We want to propose to the health department two dogs and four cats?"
>
> He said, "Yes, just put it down."
>
> I said, "All right, I'll put it down. I think we're getting off base here. This is not going to fly with them."

He said, "One more thing. What about birds?"

I said that the code says clearly, "No birds allowed in nursing homes."

He said, "But what about birds?"

I said, "What *about* birds?"

He said, "Just picture—look out your window right here. Picture that we're in January or February. We have three feet of snow outside. What sounds do you hear in the nursing home?"

I said, "Well, you hear some residents moaning. You possibly hear some laughter. You hear televisions on in different areas, maybe a little more than we'd like them to be." I said, "You'll hear an announcement over the PA system."

"What other sounds are you hearing?"

I said, "Well, you're hearing staff interacting with each other and with residents."

He said, "Yeah, but what are those sounds that are sounds of life—of positive life?"

"You're talking birdsong."

"Yes!"

I said, "How many birds are you talking to create this birdsong?"

He said, "Let's put one hundred."

"ONE HUNDRED BIRDS? IN THIS PLACE?" I said, "You've got to be out of your mind! Have you ever lived in a house that has two dogs and four cats and one hundred birds?"

And he said, "No, but wouldn't it be worth trying?"

Now that's the crux of the difference between Dr. Thomas and me.

The other three that were sitting in the room, their

eyes were bugging out of their heads now, and they were saying, "Oh my God. Do we want to do this?"

I said, "Dr. Thomas, I'm into this. I want to think outside the box. But I don't know that I want to look like a zoo, or smell like a zoo." I said, "I can't picture doing this."

He said, "Would you just hang with me?"

I said, "You've got to prove to me that this is something that has merit."

That was just the opening Thomas needed. Halbert hadn't said no. Over a few subsequent meetings, Thomas wore him and the rest of the team down. He reminded them of the Three Plagues, of the fact that people in nursing homes are dying of boredom, loneliness, and helplessness and that they wanted to find the cure for these afflictions. Wasn't anything worth trying for that?

They put the application in. It wouldn't stand a chance, Halbert figured. But Thomas took a team up to the state capital to lobby the officials in person. And they won the grant and all the regulatory waivers needed to follow through on it.

"When we got the word," Halbert recalled, "I said 'Oh my God. We're going to have to do this.'"

The job of making it work fell to Lois Greising, the director of nursing. She was in her sixties and had been working in nursing homes for years. The chance to try a new way of improving the lives of the elderly was deeply appealing to her. She told me that it felt like "this great experiment," and she decided that her task was to navigate between Thomas's sometimes oblivious optimism and the fears and inertia of the staff members.

This task was not small. Every place has a deep-seated culture as to how things are done. "Culture is the sum total of shared habits and expectations," Thomas told me. As he saw it, habits and expectations had made institutional routines and safety greater priorities than living a good life and had prevented the nursing home from successfully bringing in even one dog to live with the residents. He wanted to bring in enough animals, plants, and children to make them a regular part of every nursing home resident's life. Inevitably the settled routines of the staff would be disrupted, but then wasn't that part of the aim?

"Culture has tremendous inertia," he said. "That's why it's culture. It works because it lasts. Culture strangles innovation in the crib."

To combat the inertia, he decided they should go up against the resistance directly—"hit it hard," Thomas said. He called it the Big Bang. They wouldn't bring a dog or a cat or a bird and wait to see how everyone responded. They'd bring all the animals in more or less at once.

That fall, they moved in a greyhound named Target, a lapdog named Ginger, the four cats, and the birds. They threw out all their artificial plants and put live plants in every room. Staff members brought their kids to hang out after school; friends and family put in a garden at the back of the home and a playground for the kids. It was shock therapy.

An example of the scale: they ordered the hundred parakeets for delivery all on the same day. Had they figured out how to bring a hundred parakeets into a nursing home? No, they had not. When the delivery truck arrived, the birdcages hadn't. The driver therefore released them into the beauty salon on the ground floor, shut the door, and left. The cages arrived later that day, but in flat boxes, unassembled.

It was "total pandemonium," Thomas said. The memory of it still puts a grin on his face. He's that sort of person.

He, his wife, Jude, the nursing director, Greising, and a handful of others spent hours assembling the cages, chasing the parakeets through a cloud of feathers around the salon and delivering birds to every resident's room. The elders gathered outside the salon windows to watch.

"They laughed their butts off," Thomas said.

He marvels now at the team's incompetence. "We didn't know what the heck we were doing. *Did*, *Not*, *Know* what we were doing." Which was the beauty of it. They were so patently incompetent that most everyone dropped their guard and simply pitched in—the residents included. Whoever could do it helped line the cages with newspaper, got the dogs and the cats settled, got the kids to help out. It was a kind of glorious chaos—or, in the diplomatic words of Greising, "a heightened environment."

They had to solve numerous problems on the fly—how to feed the animals, for instance. They decided to establish daily "feeding rounds." Jude obtained an old medication cart from a decommissioned psychiatric hospital and turned it into what they called the bird-mobile. The bird-mobile was loaded up with birdseed, dog treats, and cat food, and a staff member would push it around to each room to change the newspaper liners and feed the animals. There was something beautifully subversive, Thomas said, about using a medication cart that had once dispensed metric tons of Thorazine to hand out Milk-Bones.

All sorts of crises occurred, any one of which could have ended the experiment. One night at 3:00 a.m., Thomas got a phone call from a nurse. This was not unusual. He was the medical director. But the nurse didn't want to talk to him. She wanted to talk to Jude. He put her on.

"The dog pooped on the floor," the nurse said to Jude. "Are you coming to clean it up?" As far as the nurse was concerned, this task was far below her station. She didn't go to nursing school to clean up dog crap.

Jude refused. "Complications ensued," Thomas said. The next morning, when he arrived, he found that the nurse had placed a chair over the poop, so no one would step in it, and left.

Some of the staff felt that professional animal wranglers should be hired; managing the animals wasn't a job for nursing staff and no one was paying them extra for it. In fact, they'd hardly had a raise in two or three years because of state budget cuts in nursing home reimbursements. Yet the same state government spent money on a bunch of plants and animals? Others believed that, just as in anyone's home, the animals were a responsibility that everyone should share. When you have animals, things happen, and whoever is there takes care of what needs to be done, whether it's the nursing home director or a nurse's aide. It was a battle over fundamentally different worldviews: Were they running an institution or providing a home?

Greising worked to encourage the latter view. She helped the staff balance responsibilities. Gradually people started to accept that filling Chase with life was everyone's task. And they did so not because of any rational set of arguments or compromises but because the effect on residents soon became impossible to ignore: the residents began to wake up and come to life.

"People who we had believed weren't able to speak started speaking," Thomas said. "People who had been completely withdrawn and nonambulatory started coming to the nurses' station and saying, 'I'll take the dog for a walk.' " All the parakeets were adopted and named by the residents. The lights turned back on in people's eyes. In a book he wrote about the experience, Thomas quoted from journals that the staff kept, and they described how irreplaceable the animals had become in the daily lives of residents, even ones with advanced dementia:

Gus really enjoys his birds. He listens to their singing and asks if they can have some of his coffee.

The residents are really making my job easier; many of them give me a daily report on their birds (e.g., "sings all day," "doesn't eat," "seems perkier").

M.C. went on bird rounds with me today. Usually she sits by the storage room door, watching me come and go, so this morning I asked her if she wanted to go with me. She very enthusiastically agreed, so away we went. As I was feeding and watering, M.C. held the food container for me. I explained each step to her, and when I misted the birds she laughed and laughed.

The inhabitants of Chase Memorial Nursing Home now included one hundred parakeets, four dogs, two cats, plus a colony of rabbits and a flock of laying hens. There were also hundreds of indoor plants and a thriving vegetable and flower garden. The home had on-site child care for the staff and a new after-school program.

Researchers studied the effects of this program over two years, comparing a variety of measures for Chase's residents with those of residents at another nursing home nearby. Their study found that the number of prescriptions required per resident fell to half that of the control nursing home. Psychotropic drugs for agitation, like Haldol, decreased in particular. The total drug costs fell to just 38 percent of the comparison facility. Deaths fell 15 percent.

The study couldn't say why. But Thomas thought he could. "I believe that the difference in death rates can be traced to the fundamental human need for a reason to live." And other research was consistent with this conclusion. In the early 1970s, the psychologists Judith Rodin and Ellen Langer performed an experiment in which they got a Connecticut nursing home to give each of its residents a plant. Half of them were assigned the job of watering their plant and attended a lecture on the benefits of taking on responsibilities in their lives. The other half had their plant watered for them and attended a lecture on how the staff was responsible for their well-being. After a year and a half, the group encouraged to take more responsibility—even for such a small thing as a plant—proved more active and alert and appeared to live longer.

In his book, Thomas recounted the story of a man he called Mr. L. Three months before he was admitted to the nursing home, his wife of more than sixty years died. He lost interest in eating, and his children had to help him with his daily needs more and more. Then he crashed his car into a ditch, and the police raised the possibility of its having been a suicide attempt. After Mr. L.'s discharge from the hospital, the family placed him at Chase.

Thomas recalled meeting him. "I wondered how this man had survived at all. Events of the past three months had shattered his world. He had lost his wife, his home, his freedom, and perhaps worst of all, his sense that his continued existence meant something. The joy of life was gone for him."

At the nursing home, despite antidepressant medications and efforts to encourage him, he spiraled downward. He gave up walking. He confined himself to bed. He refused to eat. Around this time, however, the new program started, and he was offered a pair of parakeets.

"He agreed, with the indifference of a person who knows he will soon be gone," Thomas said. But he began to change. "The changes were subtle at first. Mr. L. would position himself in bed so that he could watch the activities of his new charges." He began to advise the staff who came to care for his birds about what they liked and how they were doing. The birds were drawing him out. For Thomas, it was the perfect demonstration of his theory about what living things provide. In place of boredom, they offer spontaneity. In place of loneliness, they offer companionship. In place of helplessness, they offer a chance to take care of another being.

"[Mr. L.] began eating again, dressing himself, and getting out of his room," Thomas reported. "The dogs needed a walk every afternoon, and he let us know he was the man for the job." Three months later, he moved out and back into his home. Thomas is convinced the program saved his life.

Whether it did or didn't may be beside the point. The most important finding of Thomas's experiment wasn't that having a reason to live could reduce death rates for the disabled elderly. The most important finding was that it is possible to provide them with reasons to live, period. Even residents with dementia so severe that they had lost the ability to grasp much of what was going on could experience a life with greater meaning and pleasure and satisfaction. It is much harder to measure how much more worth people find in being alive than how many fewer drugs they depend on or how much longer they can live. But could anything matter more?

IN 1908, A Harvard philosopher named Josiah Royce wrote a book with the title *The Philosophy of Loyalty*. Royce was not concerned with the trials of aging. But he was concerned with a puzzle that is fundamental to anyone contemplating his or her mortality. Royce wanted to understand why simply existing— why being merely housed and fed and safe and alive—seems empty and meaningless to us. What more is it that we need in order to feel that life is worthwhile?

The answer, he believed, is that we all seek a cause beyond ourselves. This was, to him, an intrinsic human need. The cause could be large (family, country, principle) or small (a building project, the care of a pet). The important thing was that, in ascribing value to the cause and seeing it as worth making sacrifices for, we give our lives meaning.

Royce called this dedication to a cause beyond oneself loyalty. He regarded it as the opposite of individualism. The individualist puts self-interest first, seeing his own pain, pleasure, and existence as his greatest concern. For an individualist, loyalty to causes that have nothing to do with self-interest is strange. When such loyalty encourages self-sacrifice, it can even be alarming—a mistaken and irrational tendency that leaves people open to the exploitation of tyrants. Nothing could matter more than self-interest, and because when you die you are gone, self-sacrifice makes no sense.

Royce had no sympathy for the individualist view. "The selfish we had always with us," he wrote. "But the divine right to be selfish was never more ingeniously defended." In fact, he argued, human beings *need* loyalty. It does not necessarily produce happiness, and can even be painful, but we all require devotion to something more than ourselves for our lives to be endurable. Without it, we have only our desires to guide us, and they are fleeting, capricious, and insatiable. They provide, ultimately, only torment. "By nature, I am a sort of meeting place of countless streams of ancestral tendency. From moment to moment . . . I am a collection of impulses," Royce observed. "We cannot see the inner light. Let us try the outer one."

And we do. Consider the fact that we care deeply about what happens to the world after we die. If self-interest were the primary source of meaning in life, then it wouldn't matter to people if an hour after their death everyone they know were to be wiped from the face of the earth. Yet it matters greatly to most people. We feel that such an occurrence would make our lives meaningless.

The only way death is not meaningless is to see yourself as part of something greater: a family, a community, a society. If you don't, mortality is only a horror. But if you do, it is not. Loyalty, said Royce, "solves the paradox of our ordinary existence by showing us outside of ourselves the cause which is to be served, and inside of ourselves the will which delights to do this service, and which is not thwarted but enriched and expressed in such service." In more recent times, psychologists have used the term "transcendence" for a version of this idea. Above the level of self-actualization in Maslow's hierarchy of needs, they suggest the existence in people of a transcendent desire to see and help other beings achieve their potential.

As our time winds down, we all seek comfort in simple pleasures—companionship, everyday routines, the taste of good food, the warmth of sunlight on our faces. We become less interested in the rewards of achieving and accumulating, and more interested in the rewards of simply being. Yet while we may feel less ambitious, we also become concerned for our legacy. And we have a deep need to identify purposes outside ourselves that make living feel meaningful and worthwhile.

With the animals and children and plants Bill Thomas helped usher into Chase Memorial Nursing Home, a program he called the Eden Alternative, he provided a small opening for residents to express loyalty—a limited but real opportunity for them to grab on to something beyond mere existence. And they took it hungrily.

"If you're a young doc, and you bring all these animals and

children and plants into a sterile institutional nursing home circa 1992, you basically see magic happen in front of your eyes," Thomas told me. "You see people come alive. You see them begin to interact with the world, you see them begin to love and to care and to laugh. It blows your mind."

The problem with medicine and the institutions it has spawned for the care of the sick and the old is not that they have had an incorrect view of what makes life significant. The problem is that they have had almost no view at all. Medicine's focus is narrow. Medical professionals concentrate on repair of health, not sustenance of the soul. Yet—and this is the painful paradox—we have decided that they should be the ones who largely define how we live in our waning days. For more than half a century now, we have treated the trials of sickness, aging, and mortality as medical concerns. It's been an experiment in social engineering, putting our fates in the hands of people valued more for their technical prowess than for their understanding of human needs.

That experiment has failed. If safety and protection were all we sought in life, perhaps we could conclude differently. But because we seek a life of worth and purpose, and yet are routinely denied the conditions that might make it possible, there is no other way to see what modern society has done.

BILL THOMAS WANTED to remake the nursing home. Keren Wilson wanted to do away with it entirely and provide assisted living facilities instead. But they were both pursuing the same idea: to help people in a state of dependence sustain the value of existence. Thomas's first step was to give people a living being to care for; Wilson's was to give them a door they could lock and a kitchen of their own. The projects complemented each other and transformed the thinking of people involved in elder care. The question was no longer whether a better life was possible for people made dependent by physical deterioration: it was clear that it was. The question now was what the essential ingredients were. Professionals in institutions all over the world began trying to find answers. By 2010, when Lou Sanders's daughter, Shelley, went out searching for a nursing home for her father, she had no inkling of this ferment. The vast majority of places that existed for someone like him remained depressingly penitentiary. And yet new places and programs attempting to remake dependent living had begun springing up across the country and the city.

In the Boston suburbs, just twenty minutes' drive from my home, there was a new retirement community called NewBridge on the Charles. It was built on the standard continuum-of-care framework—there's independent living, assisted living, and a nursing home wing. But the nursing home that I saw on a visit not long ago looked nothing like the ones I was familiar with. Instead of housing sixty people to a floor in shared rooms along endless hospital corridors, NewBridge was divided into smaller pods housing no more than sixteen people. Each pod was called a "household" and was meant to function like one. The rooms were all private, and they were built around a common living area with a dining room, kitchen, and activity room—like a home.

The households were human size, which was a key intention. Research has found that in units with fewer than twenty people there tends to be less anxiety and depression, more socializing and friendship, an increased sense of safety, and more interaction with staff—even in cases when residents have developed dementia. But there was more to the design than just size. The households were built specifically to avoid the feel of a clinical setting. The open design let residents see what others were up to, encouraging them to join in. The presence of a central kitchen meant that, if a person felt like having a snack, he or she could go have a snack. Just standing and watching people, I could see the action spill over boundaries the way it does in real homes. Two men were playing cards in the dining room. A nurse filled out her paperwork in the kitchen instead of retreating behind a nurses' station.

There was more to the design than just architecture. The staff I met seemed to have a set of beliefs and expectations about their job that was different from what I'd encountered in other nursing homes. Walking, for instance, wasn't treated as a pathological behavior, as became instantly apparent when I met a ninety-nine-year-old great-grandmother named Rhoda Makover. Like Lou Sanders, she'd developed blood pressure problems, as well as sciatica, that resulted in frequent falls. Worse, she'd also become nearly blind from age-related retinal degeneration.

"If I see you again, I wouldn't recognize you. You're gray," Makover told me. "But you're smiling. I can see that."

Her mind remained quick and sharp. But blindness and a tendency to fall make a bad combination. It became impossible for her to live without twenty-four-hour-a-day help. In a normal nursing home, she would have been confined to a wheelchair for her safety. Here, however, she walked. Clearly there were risks. Nonetheless, the staff there understood how important mobility was—not merely for her health (in a wheelchair, her physical strength would have rapidly deteriorated) but even more for her well-being.

"Oh thank God I can go myself to the bathroom," Makover told me. "You would think it's nothing. You're young. You'll understand when you're older, but the best thing in your life is when you can go yourself to the bathroom."

She told me that in February she would turn one hundred.

"That's amazing," I said.

"That's old," she replied.

I told her my grandfather lived to almost one hundred and ten.

"God forbid," she said.

Just a few years earlier she'd had her own apartment. "I was so happy there. I was living. I was living the way people should live: I had friends, I played games. One of them would take the car, and we'd go. I was *living*." Then came the sciatica, the falls, and the loss of her vision. She was moved into a nursing home, a different one, and the experience was terrible. She lost almost everything that was her own—her furniture, her keepsakes—and found herself in a shared room, with a regimented schedule and a crucifix over her bed, "which, being Jewish, I didn't appreciate."

She was there for a year before moving to NewBridge, and it was, she said, "No comparison. *No* comparison." This was the opposite of Goffman's asylum. Human beings, the pioneers were learning, have a need for both privacy and community, for flexible daily rhythms and patterns, and for the possibility of forming caring relationships with those around them. "Here it's like living in my own home," Makover said.

Around the corner, I met Anne Braveman, seventy-nine, and Rita Kahn, eighty-six, who told me they had gone to the movies the week before. It wasn't some official, prearranged group outing. It was just two friends who decided they wanted to go see *The King's Speech* on a Thursday night. Braveman put on a nice turquoise necklace, and Kahn put on some blush, blue eye shadow, and a new outfit. A nursing assistant had to agree to join them. Braveman was paralyzed from the waist down due to multiple sclerosis and got around by motorized wheelchair; Kahn was prone to falls and needed a walker. They had to pay the $15 fare for a wheelchair-accessible vehicle to take them. But it was possible for them to go. They were looking forward to watching *Sex and the City* on DVD next.

"Have you read *Fifty Shades of Grey* yet?" Kahn asked me, impishly.

I allowed, modestly, that I had not.

"I had never heard of chains and that stuff," she said, marveling. Had I? she wanted to know.

I really didn't want to answer that.

NewBridge allowed its residents to have pets but didn't actively bring them in, the way Bill Thomas's Eden Alternative had, and so animals hadn't become a significant part of life there. But children had. NewBridge shared its grounds with a private school for students in kindergarten through eighth grade, and the two places had become deeply intertwined. Residents who didn't need significant assistance worked as tutors and school librarians. When classes studied World War II, they met with veterans who gave firsthand accounts of what they were studying in their texts. Students came in and out of New-Bridge daily, as well. The younger students held monthly events with the residents—art shows, holiday celebrations, or musical performances. Fifth and sixth graders had their fitness classes together with the residents. Middle schoolers were taught how to work with those who have dementia and took part in a buddy program with the nursing home residents. It was not unusual for children and residents to develop close individual relationships. One boy who befriended a resident with advanced Alzheimer's was even asked to speak at the man's funeral.

"Those little kids are charmers," said Rita Kahn. Her relationship with the children was one of the two most gratifying parts of her days, she told me. The other was the classes she was able to take.

"The classes! The classes! I love the classes!" She took a

current events class taught by one of the residents in indepen-dent living. When she learned that President Obama had not yet visited Israel as president, she fired off an e-mail to him.

"I really felt I had to tell this man to get off his bum and go to Israel stat."

It seemed like this kind of place might be unaffordable. But these weren't wealthy people. Rita Kahn had been a medical records administrator and her husband a high school guidance counselor. Anne Braveman had been a Massachusetts General Hospital nurse, and her husband was in the office supply busi-ness. Rhoda Makover used to be a bookkeeper and her husband a dry goods salesman. Financially, these people were no different from Lou Sanders. Indeed, 70 percent of NewBridge's nursing home residents had depleted their savings and gone onto gov-ernment assistance in order to pay for their stay.

NewBridge had been able to cultivate substantial philan-thropic support through its close ties to the Jewish community, and that had been vital to its staying afloat. But less than an hour's drive away, close to where Shelley and her husband lived, I visited a project that had nothing like NewBridge's resources and nonetheless found ways to be just as transformative. Peter Sanborn Place was built in 1983 as a subsidized apartment build-ing with seventy-three units for independent, low-income elderly people from the local community. Jacquie Carson, its director since 1996, hadn't intended to create nursing-home-level care there. But, as her tenants aged, she felt that she had to find a way to accommodate them permanently if they wanted it—and want it they did.

At first, they just needed help around their homes. Carson arranged for aides from a local agency to help with laundry, shopping, cleaning, and the like. Then some residents became weak, and she brought in physical therapists who gave them canes and walkers and taught them strengthening exercises. Some tenants required catheters, care for skin wounds, and other medical treatment. So she organized visiting nurses. When the home care agencies started telling her that she needed to move her residents into nursing homes, she remained defiant. She launched her own agency and hired people to do the job the way it should be done, giving people help with everything from meals to medical appointments.

Then one resident was diagnosed with Alzheimer's disease. "I took care of him for a couple years," Carson said, "but as he progressed, we weren't ready for that." He needed around-the-clock checks and assistance with toileting. She began to think she'd reached the limits of what she could provide and would have to put him in a nursing home. But his sons were involved with a charity, the Cure Alzheimer's Fund, which raised the money to hire Sanborn Place's first overnight staff member.

A decade or so later, just thirteen of her seventy-some residents were still independent. Twenty-five required assistance with meals, shopping, and so on. Thirty-five more required help with personal care, sometimes twenty-four hours a day. But Sanborn Place avoided becoming a certified nursing home or even an assisted living facility. Officially, it's still just a low-income apartment complex—though one with a manager who is determined to enable people to live in their own homes, in their own way, right to the end, no matter what happens.

I met a resident, Ruth Barrett, who gave me a sense of just how disabled a person could be while managing to still live in her own place. She was eighty-five and had been there eleven years, Carson said. She required oxygen, because of congestive heart failure and chronic lung disease, and she hadn't walked in four years, because of complications from arthritis and her brittle diabetes.

"I walk," Barrett objected from her motorized wheelchair.

Carson chuckled. "You don't walk, Ruthie."

"I don't walk *a lot*," Barrett replied.

Some people shrink to twigs as they age. Others become trunks. Barrett was a trunk. Carson explained that she needed twenty-four-hour assistance available and a hydraulic lift to safely move her from her wheelchair to the bed or toilet. Her memory had also faded.

"My memory is *very good*," Barrett insisted, leaning into me. Unfairly, I asked her how old she was. "Fifty-five," she said, which was off by only three decades. She remembered the past (at least the distant past) reasonably well, though. She never finished high school. She married, had a child, and divorced. She waitressed at a local diner for years to make ends meet. She eventually had three husbands in all. She mentioned one of them, and I asked her to tell me about him.

"He never killed himself working," she said.

Her desires were modest. She found comfort in her routine—a leisurely breakfast, music on the radio, a chat with friends in the lobby or her daughter on the phone, an afternoon snooze. Three or four nights a week, people gathered to watch movies on DVD in the library, and she almost always joined in. She loved going on the Friday lunch outings, even if the staff had to put her in a triple layer of Depends and clean her up when she returned. She always ordered a margarita—rocks, no salt—despite its being technically forbidden for a diabetic.

"They live like they would live in their neighborhood," Carson said of her tenants. "They still get to make poor choices for themselves if they choose."

Achieving this required more toughness than I'd realized. Carson often found herself battling the medical system. A single emergency room visit could unravel all the work she and her team had done. It was bad enough that, in the hospital, her tenants could be subject to basic medication errors, left lying on gurneys for hours (which caused their skin to break down and form open bedsores from the pressure of the thin mattresses), and assigned doctors who never called Sanborn Place for information or planning. The residents were often also shipped off to rehabilitation centers where they and their families would be told that they could never go back to apartment living again. Carson gradually worked out relationships with individual ambulance services and hospitals, which understood that Sanborn Place expected to be consulted about care for its residents and could always take them back home safely.

Even the primary care doctors the residents saw needed education. Carson recounted a conversation she'd had that day with the physician of a ninety-three-year-old woman with Alzheimer's disease.

"She's not safe," the doctor told her. "She needs to be in a nursing home."

"Why?" Carson replied. "We have bed pads. We have alarms. We have GPS tracking." The woman was well cared for. She had friends and familiar surroundings. Carson wanted him just to order some physical therapy.

"She doesn't need that. She's not going to remember how to do that," he said.

"Yes she is!" she insisted.

"She needs to be in the nursing home."

"'You need to retire,' I wanted to tell him," Carson recounted. Instead, she said to the patient, "Let's just change your doctor, because he's too old to learn." She told the woman's family, "If I'm going to waste my energy, it's not going to be on him."

I asked Carson to explain her philosophy for enabling her

residents to continue to live their own lives, whatever their condition. She said her philosophy was, "We'll figure this out."

"We will maneuver around all the obstacles there are to be maneuvered around." She spoke like a general plotting a siege. "I push probably every envelope and beyond."

The obstacles are large and small, and she was still sorting out how best to negotiate many of them. She hadn't anticipated, for example, that residents themselves might object to her efforts to help other residents stay in their homes, but some do. She said they would tell her, "So-and-so doesn't belong here anymore. She could play bingo last year. Now she doesn't even know where she is going."

Arguing with them didn't work. So Carson was now trying a new tack. "I say, 'Okay. Let's go find a place for her to live. But you're going with me, because you could be this way next year.'" So far, that has seemed enough to settle the matter.

Another example: A lot of the residents had pets, and despite the increasing difficulties they had with managing them, they wanted to keep them. So she organized her staff to empty cats' litter boxes. But the staff balked at dogs, as they required more attention than cats. Recently, though, Carson had worked out ways that her team could help with little dogs, and they'd begun allowing residents to keep them. Big dogs were still an unsolved problem. "You have to be able to take care of your dog," she said. "If your dog is running the roost, it may not be such a good idea."

Making lives meaningful in old age is new. It therefore requires more imagination and invention than making them merely safe does. The routine solutions haven't yet become well defined. So Carson and others like her are figuring them out, one person at a time. Outside the first-floor library, Ruth Beckett was chatting with a group of friends. She was a tiny ninety-year-old woman—more twig than trunk—who had been widowed years ago. She had stayed on in her house alone until a bad fall put her into a hospital and then a nursing home.

"My problem is I'm tippy," she said, "and there's no such thing as a tippy doctor."

I asked her how she'd ended up in Sanborn Place. That was when she told me about her son Wayne. Wayne was a twin born without enough oxygen. He developed cerebral palsy—he had trouble with spasticity when he walked—and was mentally delayed, as well. In adulthood, he could handle basic aspects of his life, but he needed some degree of structure and supervision. When he was in his thirties, Sanborn Place opened as a place offering just that and he was its first resident. Over the three decades since, she visited him almost every day for most of the day. But when her fall put her in a nursing home, she was no longer permitted to visit him, and he wasn't cognitively developed enough to seek to visit her. They were all but completely separated. There seemed no way around the situation. Despairing, she thought their time together was over. Carson, however, had a flash of brilliance and worked out how to take them both in. They now had apartments almost next to each other.

Just a few yards away from where I was talking with Ruth, Wayne sat in a wing chair sipping a soda and watching people come and go, his walker set to his side. They were together, as a family, again—because someone had finally understood that little mattered more to Ruth than that, not even her life.

It didn't surprise me to learn that Peter Sanborn Place had two hundred applicants on its wait list. Jacquie Carson hoped to build more capacity to accommodate them. She was, once again, trying to maneuver around all the obstacles—the lack of funding, the government bureaucracies. It will take a while, she told me. So in the meantime she's created mobile teams that can go out to help people where they live. She still wants to make it possible for everyone to live out their days wherever they can call home.

THERE ARE PEOPLE in the world who change imaginations. You can find them in the most unexpected places. And right now, in the seemingly sleepy and mundane precincts of housing for the elderly, they are cropping up all over. In eastern Massachusetts alone, I came across almost more than I could visit. I spent a couple mornings with the founders and members of Beacon Hill Villages, a kind of community cooperative in several neighborhoods of Boston dedicated to organizing affordable services—everything from plumbing repair to laundry—in order to help the elderly stay in their homes. I talked to people running assisted living homes who, against every obstacle, had stuck with the fundamental ideas Keren Wilson had planted. I've never encountered people more determined, more imaginative, and more inspiring. It depresses me to imagine how differently Alice Hobson's last years would have been if she'd been able to meet one of them—if she'd had a NewBridge, an Eden Alternative, a Peter Sanborn Place, or somewhere like them to turn to. With any of them, Alice would have had the chance to continue to be who she was despite her creeping infirmities—"to really live," as she would have put it.

The places I saw looked as different from one another as creatures in a zoo. They shared no particular shape or body parts. But the people who led them were all committed to a singular aim. They all believed that you didn't need to sacrifice your autonomy just because you needed help in your life. And I realized, in meeting these people, that they shared a very particular philosophical idea of what kind of autonomy mattered most in life.

There are different concepts of autonomy. One is autonomy as free action—living completely independently, free of coercion and limitation. This kind of freedom is a common battle cry. But it is, as Bill Thomas came to realize on his homestead in upstate New York, a fantasy—he and his wife, Jude, had two children born with severe disabilities requiring lifelong care, and someday, illness, old age, or some other mishap will leave him in need of assistance, too. Our lives are inherently dependent on others and subject to forces and circumstances well beyond our control. Having more freedom seems better than having less. But to what end? The amount of freedom you have in your life is not the measure of the worth of your life. Just as safety is an empty and even self-defeating goal to live for, so ultimately is autonomy.

The late, great philosopher Ronald Dworkin recognized that there is a second, more compelling sense of autonomy. Whatever the limits and travails we face, we want to retain the autonomy—the freedom—to be the authors of our lives. This is the very marrow of being human. As Dworkin wrote in his remarkable 1986 essay on the subject, "The value of autonomy . . . lies in the scheme of responsibility it creates: autonomy makes each of us responsible for shaping his own life according to some coherent and distinctive sense of character, conviction, and interest. It allows us to lead our own lives rather than be led along them, so that each of us can be, to the extent such a scheme of rights can make this possible, what he has made himself."

All we ask is to be allowed to remain the writers of our own story. That story is ever changing. Over the course of our lives, we may encounter unimaginable difficulties. Our concerns and desires may shift. But whatever happens, we want to retain the freedom to shape our lives in ways consistent with our character and loyalties.

This is why the betrayals of body and mind that threaten to erase our character and memory remain among our most awful tortures. The battle of being mortal is the battle to maintain the integrity of one's life—to avoid becoming so diminished or dissipated or subjugated that who you are becomes disconnected from who you were or who you want to be. Sickness and old age make the struggle hard enough. The professionals and institutions we turn to should not make it worse. But we have at last entered an era in which an increasing number of them believe their job is not to confine people's choices, in the name of safety, but to expand them, in the name of living a worthwhile life.

LOU SANDERS WAS on his way to joining the infantilized and catatonic denizens belted into the wheelchairs of a North Andover nursing home when a cousin told Shelley about a new place that had opened in the town of Chelsea, the Leonard Florence Center for Living. She should check it out, he said. It was just a short drive away. Shelley arranged for her and Lou to visit.

Lou was impressed from the first moments of the tour, when the guide mentioned something Shelley barely noted. All the rooms were single. Every nursing home Lou had ever seen had shared rooms. Losing his privacy had been among the things that scared him most. Solitude was fundamental to him. He thought he'd go crazy without it.

"My wife used to say I was a loner, but I'm not. I just like my time alone," he told me. So when the tour guide said that the Florence Center had single rooms, "I said, 'You must be kidding!'" The tour had only begun and already he was sold.

Then the guide took them through it. They called the place a

Green House. He didn't know what that meant. All he knew was, "It didn't look like a nursing home to me."

"What did it look like?" I asked.

"A home," he said.

That was the doing of Bill Thomas. After launching the Eden Alternative, he had grown restless. He was by temperament a serial entrepreneur, though without the money. He and his wife, Jude, set up a not-for-profit organization that has since taught the Eden principles to people from hundreds of nursing homes. They then became cofounders of the Pioneer Network, a kind of club for the growing number of people committed to the reinvention of elder care. It does not endorse any particular model. It simply advocates for changes that can transform our medically dominated culture of care for the elderly.

Around 2000, Thomas got a new itch. He wanted to build a home for the elderly from the ground up instead of, as he'd done in New Berlin, from the inside out. He called what he wanted to build a Green House. The plan was for it to be, as he put it, "a sheep in wolf's clothing." It needed to look to the government like a nursing home, in order to qualify for public nursing home payments, and also to cost no more than other nursing homes. It needed to have the technologies and capabilities to help people regardless of how severely disabled or impaired they might become. Yet it needed to feel to families, residents, and the people who worked there like a home, not an institution. With funding from the not-for-profit Robert Wood Johnson Foundation, he built the first Green House in Tupelo, Mississippi, in partnership with an Eden Alternative nursing home that had decided to build new units. Not long afterward, the foundation launched the National Green House Replication Initiative, which supported the construction of more than 150

Green Houses in twenty-five states—among them the Leonard Florence Center for Living that Lou had toured.

Whether it was that first home for a dozen people in a Tupelo neighborhood or the ten homes that were built in the Florence Center's six-story building, the principles have remained unchanged and echo those of other pioneers. All Green Houses are small and communal. None has more than twelve residents. At the Florence Center, the floors have two wings, each called a Green House, where about ten people live together. The residences are designed to be warm and homey—with ordinary furniture, a living room with a hearth, family-style meals around one big table, a front door with a doorbell. And they are designed to pursue the idea that a life worth living can be created, in this case, by focusing on food, homemaking, and befriending others.

It was the look of the place that attracted Lou—there was nothing dispiritingly institutional about it. But when Lou moved in, the way of life became what he valued most. He could go to bed when he wanted and wake when he wanted. Just that was a revelation to him. There was no parade of staff marching down the halls at 7:00 a.m., rustling everyone through showers and getting them dressed and wheeled into place for the pill line and group mealtime. In most nursing homes (including Chase Memorial, where Thomas had gotten his start), it had been thought that there was no other way. Efficiency demanded that the nursing aide staff have the residents ready for the cook staff, who had to have the residents ready for the activity coordination staff, who kept them out of the rooms for the cleaning staff, et cetera. So that was the way the managers designed the schedules and responsibilities. Thomas flipped the model. He took the control away from the managers and gave it to the frontline caregivers. They were each encouraged to focus on just a few residents and to become more like generalists. They did the cooking, the cleaning, and the helping with whatever need arose, whenever it arose (except for medical tasks, like giving medication, which required grabbing a nurse). As a result, they had more time and contact with each resident—time to talk, eat, play cards, whatever. Each caregiver became for people like Lou what Gerasim was for Ivan Ilyich—someone closer to a companion than a clinician.

It didn't take much to be a companion for Lou. One staff member gave him a big hug every time she saw him, and he confided to Shelley how much he loved the human contact. He had got so little of it, otherwise. On Tuesday and Thursday afternoons, he'd go down to the coffee shop and play cribbage with his friend Dave, who still visited him. Plus he'd taught the game to a man paralyzed by a stroke who lived in a home on another floor and sometimes came by Lou's place to play. An aide would hold his cards or, if necessary, Lou would, taking care not to peek. Other afternoons Shelley would come by. She'd bring the dogs, which he loved.

He was also happy, though, to spend most of the day on his own. After breakfast, he'd retreat to his room to watch television—"see about the mess," as he put it.

"I like keeping up on what's going on in politics. It's like a soap opera. Every day another chapter."

I asked him what channel he watched. Fox?

"No, MSNBC."

"MSNBC? Are you a liberal?" I said.

He grinned. "Yeah, I'm a liberal. I would vote for Dracula if he said he was a Democrat."

A while later he took some exercise, walking with his aide around the floor, or outside when the weather was good. This was a big deal to him. In his last months in assisted living, the staff had consigned him to a wheelchair, arguing it wasn't safe for him to walk, given his fainting spells. "I hated that chair," he said. The people at the Florence Center let him get rid of it and take his chances with a walker. "I'm kind of proud that I pushed the matter," he said.

He'd eat lunch at noon around the big dining table with the rest of the house. In the afternoon, if he didn't have a card game or some other plan, he'd usually read. He had subscriptions to *National Geographic* and *Newsweek*. And he still had his books. He'd finished a Robert Ludlum thriller recently. He was starting in on a book about the defeat of the Spanish Armada.

Sometimes, he pulled up to his Dell computer and surfed YouTube videos. I asked him which ones he liked to watch. He gave me an example.

"I hadn't been to China in many years"—not since the war—"so I said, let me go back to the city of Chengdu, which happens to be one of the oldest cities in the world, going back thousands of years. I was stationed near there. So I got onto the computer, and I punched in 'Chengdu.' Pretty soon I was tripping all over the city. Did you know they have synagogues there! I said 'Wow!' They tell you there's one over here, there's one over there. I was bouncing all over the place," he said. "The day goes by so fast. It goes by incredibly fast."

In the evening, after dinner, he liked to lie down on his bed, put on his headphones, and listen to music from his computer. "I like that quiet time at night. You'd be surprised. Everything is quiet. I put the easy listening on." He'd pull up Pandora and listen to smooth jazz or Benny Goodman or Spanish music—whatever he felt like. "Then I lie back and think," he said.

One time, visiting Lou, I asked him, "What makes life worth living to you?"

He paused before answering.

"I have moments when I would say I think it's time, maybe one of the days when I was at a low point," he said. "Enough is enough, you know? I would badger my Shelley. I would say, you know in Africa, when you got old and you couldn't produce anymore, they used to take you out in the jungle and leave you to be eaten by wild animals. She thought I was nuts. 'No,' I said. 'I'm not producing anything anymore. I'm costing the government money.'

"I go through that every once in a while. Then I say, 'Hey, it is what it is. Go with the flow. If they want you around, so what?'"

We had been talking in a sitting room off the kitchen with ceiling-high windows on two sides. The summer was turning to fall. The light was white and warm. We could see the town of Chelsea below us, Boston Harbor's Broad Sound in the distance, the ocean-blue sky all around. We'd been talking about the story of his life for almost two hours when it struck me that, for the first time I can remember, I did not fear reaching his phase of life. Lou was ninety-four years old and there was certainly nothing glamorous about it. His teeth were like toppled stones. He had aches in every joint. He'd lost a son and a wife, and he could no longer get around without a walker that had a yellow tennis ball jammed onto each of its front feet. He sometimes got confused and lost the thread of our conversation. But it was also apparent that he was able to live in a way that made him feel that he still had a place in this world. They still wanted him around. And that raised the possibility that the same could be the case for any of us.

The terror of sickness and old age is not merely the terror of the losses one is forced to endure but also the terror of the isolation. As people become aware of the finitude of their life, they do not ask for much. They do not seek more riches. They do not seek more power. They ask only to be permitted, insofar as possible, to keep shaping the story of their life in the world—to make choices and sustain connections to others according to their own priorities. In modern society, we have come to assume that debility and dependence rule out such autonomy. What I learned from Lou—and from Ruth Barrett, Anne Braveman, Rita Kahn, and lots of others—was that it is very much possible.

"I don't worry about the future," Lou said. "The Japanese have the word 'karma.' It means—if it's going to happen, there's nothing I can do to stop it. I know my time is limited. And so what? I've had a good shot at it."

6 · *Letting Go*

Before I began to think about what awaits my older patients—people very much like Lou Sanders and the others—I'd never ventured beyond my surgical office to follow them into their lives. But once I'd seen the transformation of elder care under way, I was struck by the simple insight on which it rested, and by its profound implications for medicine, including what happens in my own office. And the insight was that as people's capacities wane, whether through age or ill health, making their lives better often requires curbing our purely medical imperatives—resisting the urge to fiddle and fix and control. It was not hard to see how important this idea could be for the patients I encountered in my daily practice—people facing mortal circumstances at every phase of life. But it posed a difficult question: When should we try to fix and when should we not?

Sara Thomas Monopoli was just thirty-four and pregnant with her first child when the doctors at my hospital learned that she was going to die. It started with a cough and a pain in her back. Then a chest X-ray showed that her left lung had collapsed and her chest was filled with fluid. A sample of the fluid was drawn off with a long needle and sent for testing. Instead of an infection, as everyone had expected, it was lung cancer, and it had already spread to the lining of her chest. Her pregnancy was thirty-nine weeks along, and the obstetrician who had ordered the test broke the news to her as she sat with her husband and her parents. The obstetrician didn't get into the prognosis—she would bring in an oncologist for that—but Sara was stunned. Her mother, who had lost her best friend to lung cancer, began crying.

The doctors wanted to start treatment right away, and that meant inducing labor to get the baby out. For the moment, though, Sara and her husband, Rich, sat by themselves on a quiet terrace off the labor floor. It was a warm Monday in June. She took Rich's hands, and they tried to absorb what they had heard. She had never smoked or lived with anyone who had. She exercised. She ate well. The diagnosis was bewildering. "This is going to be okay," Rich told her. "We're going to work through this. It's going to be hard, yes. But we'll figure it out. We can find the right treatment." For the moment, however, they had a baby to think about.

"So Sara and I looked at each other," Rich recalled, "and we said, 'We don't have cancer on Tuesday. It's a cancer-free day. We're having a baby. It's exciting. And we're going to enjoy our baby.' " On Tuesday, at 8:55 p.m., Vivian Monopoli, seven pounds nine ounces, was born. She had wavy brown hair, like her mom, and she was in perfect health.

The next day, Sara underwent blood tests and body scans. Paul Marcoux, an oncologist, met with her and her family to discuss the findings. He explained that she had a non-small cell lung cancer that had started in her left lung. Nothing she had done had brought the disease on. More than 15 percent of lung cancers—more than people realize—occur in nonsmokers. Hers was advanced, having metastasized to multiple lymph nodes in her chest and its lining. The cancer was inoperable. But there were chemotherapy options, notably a drug called erlotinib, which targets a gene mutation commonly found in lung cancers of female nonsmokers; 85 percent of them respond to the drug, and, as Marcoux said, "some of these responses can be long-term."

Words like "respond" and "long-term" provide a reassuring gloss on a dire reality. There is no cure for lung cancer at this stage. Even with chemotherapy, the median survival is about a year. But it seemed harsh and pointless for him to confront Sara and Rich with that fact now. Vivian was in a bassinet by the bed. They were working hard to be optimistic. As Sara and Rich later told the social worker who was sent to see them, they did not want to focus on survival statistics. They wanted to focus on "aggressively managing" this diagnosis.

So Sara started on the erlotinib, which produced an itchy, acne-like facial rash and numbing tiredness. She also underwent a needle drainage of the fluid around her lung, but the fluid kept coming back and the painful procedure had to be repeated again and again. So a thoracic surgeon was called in to place a small permanent tube in her chest, which she could drain by turning a stopcock whenever fluid accumulated and interfered with her breathing. Three weeks after her childbirth, she was readmitted to the hospital with severe shortness of breath from a pulmonary embolism—a blood clot in an artery to the lungs, which is dangerous but not uncommon in cancer patients. She was started on a blood thinner. Then test results showed that her tumor cells did not have the mutation that erlotinib targets. When Marcoux told Sara that the drug wasn't going to work, she had an almost violent physical reaction to the news, bolting to the bathroom in mid-discussion with a sudden bout of diarrhea.

Marcoux recommended a different, more standard chemotherapy, with two drugs called carboplatin and paclitaxel. But the paclitaxel triggered an extreme, nearly overwhelming allergic response, so he switched her to a regimen of carboplatin plus gemcitabine. Response rates, he said, were still very good for patients on this therapy.

She spent the remainder of the summer at home, with Vivian and her husband and her parents, who had moved in to help. She loved being a mother. Between chemotherapy cycles, she began trying to get her life back.

Then, in October, a CT scan showed that the tumor deposits in her left chest and in her lymph nodes had grown substantially. The chemotherapy had failed. She was switched to a drug called pemetrexed. Studies had shown that it could produce markedly longer survival in some patients. In reality, only a small percentage of patients gained very much. On average, the drug extended survival by only two months—from eleven to thirteen months—and that was in patients who, unlike Sara, had responded to first-line chemotherapy.

She worked hard to take the setbacks and side effects in stride. She was upbeat by nature, and she managed to maintain her optimism. Little by little, however, she grew sicker—increasingly exhausted and short of breath. In a matter of months, it was as if she'd aged decades. By November, she didn't have the wind to walk the length of the hallway from the parking garage to Marcoux's office; Rich had to push her in a wheelchair.

A few days before Thanksgiving, she had another CT scan, which showed that the pemetrexed—her third drug regimen—wasn't working, either. The lung cancer had spread: from the left chest to the right, to the liver, to the lining of her abdomen, and to her spine. Time was running out.

THIS IS THE moment in Sara's story that poses our difficult question, one for everyone living in our era of modern medicine: What do we want Sara and her doctors to do now? Or, to put it another way, if you were the one who had metastatic cancer—or, for that matter, any similarly advanced and incurable condition—what would you want your doctors to do?

The issue has gotten attention, in recent years, for reasons of expense. The soaring cost of health care has become the greatest threat to the long-term solvency of most advanced nations, and the incurable account for a lot of it. In the United States, 25 percent of all Medicare spending is for the 5 percent of patients who are in their final year of life, and most of that money goes for care in their last couple of months that is of little apparent benefit. The United States is often thought to be unusual in this regard, but it doesn't appear to be. Data from elsewhere are more limited, but where they are available—for instance, from countries like the Netherlands and Switzerland—the results are similar.

Spending on a disease like cancer tends to follow a particular pattern. There are high initial costs as the cancer is treated, and then, if all goes well, these costs taper off. A 2011 study, for instance, found that medical spending for a breast cancer patient in the first year of diagnosis averaged an estimated $28,000, the vast majority of it for the initial diagnostic testing, surgery, and, where necessary, radiation and chemotherapy. Costs fell after that to about $2,000 a year. For a patient whose cancer proves fatal, though, the cost curve is U-shaped, rising toward the end—to an average of $94,000 during the last year of life with a metastatic breast cancer. Our medical system is excellent at trying to stave off death with $12,000-a-month chemotherapy,

$4,000-a-day intensive care, $7,000-an-hour surgery. But, ultimately, death comes, and few are good at knowing when to stop.

While seeing a patient in an intensive care unit at my hospital, I stopped to talk with the critical care physician on duty, someone I'd known since college. "I'm running a warehouse for the dying," she said bleakly. Of the ten patients in her unit, she said, only two were likely to leave the hospital for any length of time. More typical was an almost eighty-year-old woman at the end of her life, with irreversible congestive heart failure, who was in the ICU for the second time in three weeks, drugged to oblivion and tubed in most natural orifices as well as a few artificial ones. Or the seventy-year-old with a cancer that had metastasized to her lungs and bone and a fungal pneumonia that arises only in the final phase of the illness. She had chosen to forgo treatment, but her oncologist pushed her to change her mind, and she was put on a ventilator and antibiotics. Another woman, in her eighties, with end-stage respiratory and kidney failure, had been in the unit for two weeks. Her husband had died after a long illness, with a feeding tube and a tracheostomy, and she had mentioned that she didn't want to die that way. But her children couldn't let her go and asked to proceed with the placement of various devices: a permanent tracheostomy, a feeding tube, and a dialysis catheter. So now she just lay there tethered to her pumps, drifting in and out of consciousness.

Almost all these patients had known, for some time, that they had a terminal condition. Yet they—along with their families and doctors—were unprepared for the final stage.

"We are having more conversation now about what patients want for the end of their life, by far, than they have had in all their lives to this point," my friend said. "The problem is that's way too late."

In 2008, the national Coping with Cancer project published a study showing that terminally ill cancer patients who were put on a mechanical ventilator, given electrical defibrillation or chest compressions, or admitted, near death, to intensive care had a substantially worse quality of life in their last week than those who received no such interventions. And, six months after their death, their caregivers were three times as likely to suffer major depression. Spending one's final days in an ICU because of terminal illness is for most people a kind of failure. You lie attached to a ventilator, your every organ shutting down, your mind teetering on delirium and permanently beyond realizing that you will never leave this borrowed, fluorescent place. The end comes with no chance for you to have said good-bye or "It's okay" or "I'm sorry" or "I love you."

People with serious illness have priorities besides simply prolonging their lives. Surveys find that their top concerns include avoiding suffering, strengthening relationships with family and friends, being mentally aware, not being a burden on others, and achieving a sense that their life is complete. Our system of technological medical care has utterly failed to meet these needs, and the cost of this failure is measured in far more than dollars. The question therefore is not how we can afford this system's expense. It is how we can build a health care system that will actually help people achieve what's most important to them at the end of their lives.

IN THE PAST, when dying was typically a more precipitous process, we did not have to think about a question like this. Though some diseases and conditions had a drawn-out natural history—tuberculosis is the classic example—without the intervention of modern medicine, with its scans to diagnose problems early and its treatments to extend life, the interval between recognizing that you had a life-threatening ailment and dying was commonly a matter of days or weeks. Consider how our presidents died before the modern era. George Washington developed a throat infection at home on December 13, 1799, that killed him by the next evening. John Quincy Adams, Millard Fillmore, and Andrew Johnson all succumbed to strokes and died within two days. Rutherford Hayes had a heart attack and died three days later. Others did have a longer course: James Monroe and Andrew Jackson died from progressive and far longer-lasting (and highly dreaded) tubercular consumption. Ulysses Grant's oral cancer took a year to kill him. But, as end-of-life researcher Joanne Lynn has observed, people generally experienced life-threatening illness the way they experienced bad weather—as something that struck with little warning. And you either got through it or you didn't.

Dying used to be accompanied by a prescribed set of customs. Guides to *ars moriendi*, the art of dying, were extraordinarily popular; a medieval version published in Latin in 1415 was reprinted in more than a hundred editions across Europe. People believed death should be accepted stoically, without fear or self-pity or hope for anything more than the forgiveness of God. Reaffirming one's faith, repenting one's sins, and letting go of one's worldly possessions and desires were crucial, and the guides provided families with prayers and questions for the dying in order to put them in the right frame of mind during their final hours. Last words came to hold a particular place of reverence.

These days, swift catastrophic illness is the exception. For most people, death comes only after long medical struggle with an ultimately unstoppable condition—advanced cancer, dementia, Parkinson's disease, progressive organ failure (most commonly the heart, followed in frequency by lungs, kidneys, liver), or else just the accumulating debilities of very old age. In all such cases, death is certain, but the timing isn't. So everyone struggles with this uncertainty—with how, and when, to accept that the battle is lost. As for last words, they hardly seem to exist anymore. Technology can sustain our organs until we are well past the point of awareness and coherence. Besides, how do you attend to the thoughts and concerns of the dying when medicine has made it almost impossible to be sure who the dying even are? Is someone with terminal cancer, dementia, or incurable heart failure dying, exactly?

I was once the surgeon for a woman in her sixties who had severe chest and abdominal pain from a bowel obstruction that had ruptured her colon, caused her to have a heart attack, and put her into septic shock and kidney failure. I performed an emergency operation to remove the damaged length of colon and give her a colostomy. A cardiologist stented open her coronary arteries. We put her on dialysis, a ventilator, and intravenous feeding, and she stabilized. After a couple of weeks, though, it was clear that she was not going to get much better. The septic shock had left her with heart and respiratory failure as well as dry gangrene of her foot, which would have to be amputated. She had a large, open abdominal wound with leaking bowel contents, which would require weeks of twice-a-day dressing changes and cleansing in order to heal. She would not be able to eat. She would need a tracheostomy. Her kidneys were gone, and she would have to spend three days a week on a dialysis machine for the rest of her life.

She was unmarried and without children. So I sat with her sisters in the ICU's family room to talk about whether we should proceed with the amputation and the tracheostomy.

"Is she dying?" one of the sisters asked me.

I didn't know how to answer the question. I wasn't even

sure what the word "dying" meant anymore. In the past few decades, medical science has rendered obsolete centuries of experience, tradition, and language about our mortality and created a new difficulty for mankind: how to die.

ONE SPRING FRIDAY morning, I went on patient rounds with Sarah Creed, a nurse with the hospice service that my hospital system operated. I didn't know much about hospice. I knew that it specialized in providing "comfort care" for the terminally ill, sometimes in special facilities, though nowadays usually at home. I knew that, in order for a patient of mine to be eligible, I had to write a note certifying that he or she had a life expectancy of less than six months. I also knew few patients who had chosen it, except in their very last few days, because they had to sign a form indicating that they understood their disease was terminal and that they were giving up on medical care that aimed to stop it. The picture I had of hospice was of a morphine drip. It was not of this brown-haired and blue-eyed former ICU nurse with a stethoscope, knocking on Lee Cox's door on a quiet morning in Boston's Mattapan neighborhood.

"Hi, Lee," Creed said when she entered the house.

"Hi, Sarah," Cox said. She was seventy-two years old. She'd had several years of declining health due to congestive heart failure from a heart attack and pulmonary fibrosis, a progressive and irreversible lung disease. Doctors tried slowing the disease with steroids, but they didn't work. She had cycled in and out of the hospital, each time in worse shape. Ultimately, she accepted hospice care and moved in with her niece for support. She was dependent on oxygen and unable to do the most ordinary tasks. Just answering the door, with her thirty-foot length of oxygen tubing trailing after her, had left her winded. She stood resting for a moment, her lips pursed and her chest heaving.

Creed took Cox's arm gently as we walked to the kitchen to sit down, asking her how she had been doing. Then she asked a series of questions, targeting issues that tend to arise in patients with terminal illness. Did Cox have pain? How was her appetite, thirst, sleeping? Any trouble with confusion, anxiety, or restlessness? Had her shortness of breath grown worse? Was there chest pain or heart palpitations? Abdominal discomfort? Trouble with constipation or urination or walking?

She did have some new troubles. When she walked from the bedroom to the bathroom, she said, it now took at least five minutes to catch her breath, and that frightened her. She was also getting chest pain. Creed pulled a blood pressure cuff from her medical bag. Cox's blood pressure was acceptable, but her heart rate was high. Creed listened to her heart, which had a normal rhythm, and to her lungs, hearing the fine crackles of her pulmonary fibrosis but also a new wheeze. Her ankles were swollen with fluid, and when Creed asked for her pillbox she saw that Cox was out of her heart medication. She asked to see Cox's oxygen equipment. The liquid-oxygen cylinder at the foot of her neatly made bed was filled and working properly. The nebulizer equipment for her inhaler treatments, however, was broken.

Given the lack of heart medication and inhaler treatments, it was no wonder that she had worsened. Creed called Cox's pharmacy. They said that her refills had been waiting all along. So Creed contacted Cox's niece to pick up the medicine when she came home from work. She also called the nebulizer supplier for same-day emergency service.

She then chatted with Cox in the kitchen for a few minutes. Cox's spirits were low. Creed took her hand. Everything was going to be all right, she said. She reminded her about the good days she'd had—the previous weekend, for example, when she'd been able to go out with her portable oxygen cylinder to shop with her niece and get her hair colored.

I asked Cox about her earlier life. She had made radios in a Boston factory. She and her husband had had two children and several grandchildren.

When I asked her why she had chosen hospice care, she looked downcast. "The lung doctor and heart doctor said they couldn't help me anymore," she said. Creed glared at me. My questions had made Cox sad again.

She told a story of the trials of aging overlain with the trials of having an illness that she knew would someday claim her. "It's good to have my niece and her husband helping to watch me every day," she said. "But it's not my home. I feel like I'm in the way." Multigenerational living fell short of its nostalgic image, again.

Creed gave her a hug and one last reminder before we left. "What do you do if you have chest pain that doesn't go away?" she asked.

"Take a nitro," Cox said, referring to the nitroglycerin pill that she can slip under her tongue.

"And?"

"Call you."

"Where's the number?"

She pointed to the twenty-four-hour hospice call number that was taped beside her phone.

Outside, I confessed that I was confused by what Creed was doing. A lot of it seemed to be about extending Cox's life. Wasn't the goal of hospice to let nature take its course?

"That's not the goal," Creed said. The difference between standard medical care and hospice is not the difference between treating and doing nothing, she explained. The difference was in the priorities. In ordinary medicine, the goal is to extend life. We'll sacrifice the quality of your existence now—by performing surgery, providing chemotherapy, putting you in intensive care—for the chance of gaining time later. Hospice deploys nurses, doctors, chaplains, and social workers to help people with a fatal illness have the fullest possible lives right now—much as nursing home reformers deploy staff to help people with severe disabilities. In terminal illness that means focusing on objectives like freedom from pain and discomfort, or maintaining mental awareness for as long as feasible, or getting out with family once in a while—not on whether Cox's life would be longer or shorter. Nonetheless, when she was transferred to hospice care, her doctors thought that she wouldn't live much longer than a few weeks. With the supportive hospice therapy she received, she had already lived for a year.

Hospice is not an easy choice for a person to make. A hospice nurse enters people's lives at a strange moment—when they have understood that they have a fatal illness but not necessarily acknowledged that they are dying. "I'd say only about a quarter have accepted their fate when they come into hospice," Creed said. When she first encounters her patients, many feel that their doctors have simply abandoned them. "Ninety-nine percent understand they're dying, but one hundred percent hope they're not," she told me. "They still want to beat their disease." The initial visit is always tricky, but she has found ways to smooth things over. "A nurse has five seconds to make a patient like you and trust you. It's in the whole way you present yourself. I do not come in saying, 'I'm so sorry.' Instead, it's: 'I'm the hospice nurse, and here's what I have to offer you to make your life better. And I know we don't have a lot of time to waste.'"

That was how she started with Dave Galloway, whom we

visited after leaving Lee Cox's home. He was forty-two years old. He and his wife, Sharon, were both Boston firefighters. They had a three-year-old daughter. He had pancreatic cancer, which had spread; his upper abdomen was now solid with tumor. During the past few months, the pain had often become unbearable, and he was admitted to the hospital several times for pain crises. At his most recent admission, about a week earlier, it was found that the tumor had perforated his intestine. There wasn't even a temporary fix for this problem. The medical team started him on intravenous nutrition and offered him a choice between going to the intensive care unit and going home with hospice. He chose to go home.

"I wish we'd gotten involved sooner," Creed told me. When she and the hospice's supervising doctor, JoAnne Nowak, evaluated Galloway upon his arrival at home, he appeared to have only a few days left. His eyes were hollow. His breathing was labored. Fluid swelled his entire lower body to the point that his skin blistered and wept. He was almost delirious with abdominal pain.

They got to work. They set up a pain pump with a button that let him dispense higher doses of narcotic than he had been allowed. They arranged for an electric hospital bed, so that he could sleep with his back raised. They also taught Sharon how to keep Dave clean, protect his skin from breakdown, and handle the crises to come. Creed told me that part of her job is to take the measure of a patient's family, and Sharon struck her as unusually capable. She was determined to take care of her husband to the end, and perhaps because she was a firefighter, she had the resilience and the competence to do so. She did not want to hire a private-duty nurse. She handled everything, from the IV lines and the bed linens to orchestrating family members to lend a hand when she needed help.

Creed arranged for a specialized "comfort pack" to be delivered by FedEx and stored in a minirefrigerator by Dave's bed. It contained a dose of morphine for breakthrough pain or shortness of breath, Ativan for anxiety attacks, Compazine for nausea, Haldol for delirium, Tylenol for fever, and atropine for drying up the wet upper-airway rattle that people can get in their final hours. If any such problem developed, Sharon was instructed to call the twenty-four-hour hospice nurse on duty, who would provide instructions about which rescue medications to use and, if necessary, come out to help.

Dave and Sharon were finally able to sleep through the night at home. Creed or another nurse came to see him every day, sometimes twice a day. Three times that week, Sharon used the emergency hospice line to help her deal with Dave's pain crises or hallucinations. After a few days, they were even able to go out to a favorite restaurant; he wasn't hungry, but they enjoyed just being there and the memories it stirred.

The hardest part so far, Sharon said, was deciding to forgo the two-liter intravenous feedings that Dave had been receiving each day. Although they were his only source of calories, the hospice staff encouraged discontinuing them because his body did not seem to be absorbing the nutrition. The infusion of sugars, proteins, and fats made the painful swelling of his skin and his shortness of breath worse—and for what? The mantra was: live for now. Sharon had balked, for fear that she'd be starving him. The night before our visit, however, she and Dave decided to try going without the infusion. By morning, the swelling was markedly reduced. He could move more, and with less discomfort. He also began to eat a few morsels of food, just for the taste of it, and that made Sharon feel better about the decision.

When we arrived, Dave was making his way back to bed

after a shower, his arm around his wife's shoulders and his slippered feet taking one shuffling step at a time.

"There's nothing he likes better than a long, hot shower," Sharon said. "He'd live in the shower if he could."

Dave sat on the edge of his bed in fresh pajamas, catching his breath, and Creed spoke to him as his daughter, Ashlee, ran in and out of the room in her beaded pigtails, depositing stuffed animals in her dad's lap.

"How's your pain on a scale of one to ten?" Creed asked.

"A six," he said.

"Did you hit the pump?"

He didn't answer for a moment. "I'm reluctant," he admitted.

"Why?" Creed asked.

"It feels like defeat," he said.

"Defeat?"

"I don't want to become a drug addict," he explained. "I don't want to need this."

Creed got down on her knees in front of him. "Dave, I don't know anyone who can manage this kind of pain without the medication," she said. "It's not defeat. You've got a beautiful wife and daughter, and you're not going to be able to enjoy them with the pain."

"You're right about that," he said, looking at Ashlee as she gave him a little horse. And he pressed the button.

Dave Galloway died one week later—at home, at peace, and surrounded by family. A week after that, Lee Cox died, too. But as if to show just how resistant to formula human lives are, Cox had never reconciled herself to the incurability of her illnesses. So when her family found her in cardiac arrest one morning, they followed her wishes and called 911 instead of the hospice service. The emergency medical technicians and firefighters and police rushed in. They pulled off her clothes and pumped her chest, put a tube in her airway and forced oxygen into her lungs, and tried to see if they could shock her heart back. But such efforts rarely succeed with terminal patients, and they did not succeed with her.

Hospice has tried to offer a new ideal for how we die. Although not everyone has embraced its rituals, those who have are helping to negotiate an *ars moriendi* for our age. But doing so represents a struggle—not only against suffering but also against the seemingly unstoppable momentum of medical treatment.

JUST BEFORE THANKSGIVING, Sara Monopoli, her husband, Rich, and her mother, Dawn Thomas, met with Dr. Marcoux to discuss the options she had left. By this point, Sara had undergone three rounds of chemotherapy with limited, if any, effect. Perhaps Marcoux could have discussed what she most wanted as death neared and how best to achieve those wishes. But the signal he got from Sara and her family was that they wished to talk only about the next treatment options. They did not want to talk about dying.

Later, after her death, I spoke to Sara's husband and her parents. Sara knew that her disease was incurable, they pointed out. The week after she was given the diagnosis and delivered her baby, she spelled out her wishes for Vivian's upbringing after she was gone. On several occasions, she told her family that she did not want to die in the hospital. She wanted to spend her final moments peacefully at home. But the prospect that those moments might be coming soon, that there might be no way to slow the disease, "was not something she or I wanted to discuss," her mother said.

Her father, Gary, and her twin sister, Emily, still held out hope for a cure. The doctors simply weren't looking hard enough, they felt. "I just couldn't believe there wasn't something," Gary said. For Rich, the experience of Sara's illness had been disorienting: "We had a baby. We were young. And this was so shocking and so odd. We never discussed stopping treatment."

Marcoux took the measure of the room. With almost two decades of experience treating lung cancer, he had been through many of these conversations. He has a calm, reassuring air and a native Minnesotan's tendency to avoid confrontation or over-intimacy. He tries to be scientific about decisions.

"I know that the vast majority of my patients are going to die of their disease," he told me. The data show that, after fail-ure of second-line chemotherapy, lung cancer patients rarely get any added survival time from further treatments and often suf-fer significant side effects. But he, too, has his hopes.

He told them that, at some point, "supportive care" was an option for them to think about. But, he went on, there were also experimental therapies. He told them about several that were under trial. The most promising was a Pfizer drug that targeted one of the mutations found in her cancer's cells. Sara and her family instantly pinned their hopes on it. The drug was so new that it didn't even have a name, just a number—PF0231006— and this made it all the more enticing.

There were a few hovering issues, including the fact that the scientists didn't yet know the safe dose. The drug was only in a Phase I trial—that is, a trial designed to determine the toxicity of a range of doses, not whether the drug worked. Furthermore, a test of the drug against her cancer cells in a petri dish showed no effect. But Marcoux thought that these were not decisive obstacles, just negatives. The critical problem was that the rules of the trial excluded Sara because of the pulmonary embolism she had developed that summer. To enroll, she would need to wait two months in order to get far enough past the episode. In the meantime, he suggested trying another conventional chemo-therapy, called vinorelbine. Sara began the treatment the Monday after Thanksgiving.

It's worth pausing to consider what had just happened. Step by step, Sara ended up on a fourth round of chemotherapy, one with a minuscule likelihood of altering the course of her disease and a great likelihood of causing debilitating side effects. An opportunity to prepare for the inevitable was forgone. And it all happened because of an assuredly normal circumstance: a patient and family unready to confront the reality of her disease.

I asked Marcoux what he hopes to accomplish for terminal lung cancer patients when they first come to see him. "I'm thinking, can I get them a pretty good year or two out of this?" he said. "Those are my expectations. For me, the long tail for a patient like her is three to four years." But this is not what people want to hear. "They're thinking ten to twenty years. You hear that time and time again. And I'd be the same way if I were in their shoes."

You'd think doctors would be well equipped to navigate the shoals here, but at least two things get in the way. First, our own views may be unrealistic. A study led by the sociologist Nicholas Christakis asked the doctors of almost five hundred terminally ill patients to estimate how long they thought their patient would survive and then followed the patients. Sixty-three percent of doctors overestimated their patient's survival time. Just 17 percent underestimated it. The average estimate was 530 percent too high. And the better the doctors knew their patients, the more likely they were to err.

Second, we often avoid voicing even these sentiments. Studies find that although doctors usually tell patients when a cancer is not curable, most are reluctant to give a specific prognosis, even when pressed. More than 40 percent of oncologists admit to offering treatments that they believe are unlikely to work. In an era in which the relationship between patient and doctor is increasingly miscast in retail terms—"the customer is always right"—doctors are especially hesitant to trample on a patient's expectations. You worry far more about being overly pessimistic than you do about being overly optimistic. And talking about dying is enormously fraught. When you have a patient like Sara Monopoli, the last thing you want to do is grapple with the truth. I know, because Marcoux wasn't the only one avoiding that conversation with her. I was, too.

Earlier that summer, a PET scan had revealed that, in addition to her lung cancer, she had thyroid cancer, which had spread to the lymph nodes of her neck, and I was called in to decide whether to operate. This second, unrelated cancer was in fact operable. But thyroid cancers take years to become lethal. Her lung cancer would almost certainly end her life long before her thyroid cancer caused any trouble. Given the extent of the surgery that would have been required and the potential complications, the best course was to do nothing. But explaining my reasoning to Sara meant confronting the mortality of her lung cancer, something that I felt ill prepared to do.

Sitting in my clinic, Sara did not seem discouraged by the discovery of this second cancer. She seemed determined. She'd read about the good outcomes from thyroid cancer treatment. So she was geared up, eager to discuss when to operate. And I found myself swept along by her optimism. Suppose I was wrong, I wondered, and she proved to be that miracle patient who survived metastatic lung cancer? How could I let her thyroid cancer go untreated?

My solution was to avoid the subject altogether. I told Sara that there was relatively good news about her thyroid cancer—it was slow growing and treatable. But the priority was her lung cancer, I said. Let's not hold up the treatment for that. We could monitor the thyroid cancer for now and plan surgery in a few months.

I saw her every six weeks and noted her physical decline from one visit to the next. Yet, even in a wheelchair, Sara would always arrive smiling, makeup on and bangs bobby-pinned out of her eyes. She'd find small things to laugh about, like the strange protuberances the tubes made under her dress. She was ready to try anything, and I found myself focusing on the news about experimental therapies for her lung cancer. After one of her chemotherapies seemed to shrink the thyroid cancer slightly, I even raised with her the possibility that an experimental therapy could work against both her cancers, which was sheer fantasy. Discussing a fantasy was easier—less emotional, less explosive, less prone to misunderstanding—than discussing what was happening before my eyes.

Between the lung cancer and the chemo, Sara became steadily sicker. She slept most of the time and could do little out of the house. Clinic notes from December describe shortness of breath, dry heaves, coughing up blood, severe fatigue. In addition to the drainage tubes in her chest, she required needle-drainage procedures in her abdomen every week or two to relieve the severe pressure from the liters of fluid that the cancer was producing there.

A CT scan in December showed that the lung cancer was spreading through her spine, liver, and lungs. When we met in January, she could move only slowly and uncomfortably. Her lower body had become so swollen that her skin was taut. She couldn't speak more than a sentence without pausing for breath. By the first week of February, she needed oxygen at home to breathe. Enough time had elapsed since her pulmonary embolism, however, that she could start on Pfizer's experimental drug.

She just needed one more set of scans for clearance. These revealed that the cancer had spread to her brain, with at least nine metastatic growths up to half an inch in size scattered across both hemispheres. The experimental drug was not designed to cross the blood-brain barrier. PF0231006 was not going to work.

And still Sara, her family, and her medical team remained in battle mode. Within twenty-four hours, Sara was brought in to see a radiation oncologist for whole-brain radiation to try to reduce the metastases. On February 12, she completed five days of radiation treatment, which left her immeasurably fatigued, barely able to get out of bed. She ate almost nothing. She weighed twenty-five pounds less than she had in the fall. She confessed to Rich that, for the past two months, she had experienced double vision and was unable to feel her hands.

"Why didn't you tell anyone?" he asked her.

"I just didn't want to stop treatment," she said. "They would make me stop."

She was given two weeks to recover her strength after the radiation. Then we had a different experimental drug she could try, one from a small biotech company. She was scheduled to start on February 25. Her chances were rapidly dwindling. But who was to say they were zero?

In 1985, the paleontologist and writer Stephen Jay Gould published an extraordinary essay entitled "The Median Isn't the Message" after he had been given a diagnosis, three years earlier, of abdominal mesothelioma, a rare and lethal cancer usually associated with asbestos exposure. He went to a medical library when he got the diagnosis and pulled out the latest scientific articles on the disease. "The literature couldn't have been more brutally clear: mesothelioma is incurable, with a median survival of only eight months after discovery," he wrote. The news was devastating. But then he began looking at the graphs of the patient-survival curves.

Gould was a naturalist and more inclined to notice the variation around the curve's middle point than the middle point itself. What the naturalist saw was remarkable variation. The patients were not clustered around the median survival but, instead, fanned out in both directions. Moreover, the curve was skewed to the right, with a long tail, however slender, of patients who lived many years longer than the eight-month median. This is where he found solace. He could imagine himself surviving far out along that long tail. And survive he did. Following surgery and experimental chemotherapy, he lived twenty more years before dying, in 2002, at the age of sixty, from a lung cancer unrelated to his original disease.

"It has become, in my view, a bit too trendy to regard the acceptance of death as something tantamount to intrinsic dignity," he wrote in his 1985 essay. "Of course I agree with the preacher of Ecclesiastes that there is a time to love and a time to die—and when my skein runs out I hope to face the end calmly and in my own way. For most situations, however, I prefer the more martial view that death is the ultimate enemy—and I find nothing reproachable in those who rage mightily against the dying of the light."

I think of Gould and his essay every time I have a patient with a terminal illness. There is almost always a long tail of possibility, however thin. What's wrong with looking for it? Nothing, it seems to me, unless it means we have failed to prepare for the outcome that's vastly more probable. The trouble is that we've built our medical system and culture around the long tail. We've created a multitrillion-dollar edifice for dispensing the medical equivalent of lottery tickets—and have only the rudiments of a system to prepare patients for the near certainty that those tickets will not win. Hope is not a plan, but hope is our plan.

FOR SARA, THERE would be no miraculous recovery, and when the end approached, neither she nor her family was prepared. "I always wanted to respect her request to die peacefully at home," Rich later told me. "But I didn't believe we could make it happen. I didn't know how."

On the morning of Friday, February 22, three days before she was to start her new round of chemo, Rich awoke to find his wife sitting upright beside him, pitched forward on her arms, eyes wide, struggling for air. She was gray, breathing fast, her body heaving with each open-mouthed gasp. She looked as if she were drowning. He tried turning up the oxygen in her nasal tubing, but she got no better.

"I can't do this," she said, pausing between each word. "I'm scared."

He had no emergency kit in the refrigerator. No hospice nurse to call. And how was he to know whether this new development was fixable?

We'll go to the hospital, he told her. When he asked if they should drive, she shook her head, so he called 911 and told her mother, Dawn, who was in the next room, what was going on. A few minutes later, firemen swarmed up the stairs to her bedroom, sirens wailing outside. As they lifted Sara into the ambulance on a stretcher, Dawn came out in tears.

"We're going to get ahold of this," Rich told her. This was just another trip to the hospital, he said to himself. The doctors would figure out how to fix her.

At the hospital, Sara was diagnosed with pneumonia. That troubled the family because they thought they'd done everything to keep infection at bay. They'd washed hands scrupulously, limited visits by people with young children, even limited Sara's time with baby Vivian if she showed the slightest sign of a runny nose. But Sara's immune system and her ability to clear her lung secretions had been steadily weakened by the rounds of radiation and chemotherapy as well as by the cancer.

In another way, the diagnosis of pneumonia was reassuring, because it was just an infection. It could be treated. The medical team started Sara on intravenous antibiotics and high-flow oxygen through a mask. The family gathered at her bedside, hoping for the antibiotics to work. The problem could be reversible, they told one another. But that night and the next morning her breathing only grew more labored.

"I can't think of a single funny thing to say," Emily told Sara as their parents looked on.

"Neither can I," Sara murmured. Only later did the family realize that those were the last words they would ever hear from her. After that, she began to drift in and out of consciousness. The medical team had only one option left: to put her on a ventilator. Sara was a fighter, right? And the next step for fighters is to escalate to intensive care.

THIS IS A modern tragedy, replayed millions of times over. When there is no way of knowing exactly how long our skeins will run—and when we imagine ourselves to have much more time than we do—our every impulse is to fight, to die with chemo in our veins or a tube in our throats or fresh sutures in our flesh. The fact that we may be shortening or worsening the time we have left hardly seems to register. We imagine that we can wait until the doctors tell us that there is nothing more they can do. But rarely is there nothing more that doctors can do. They can give toxic drugs of unknown efficacy, operate to try to remove part of the tumor, put in a feeding tube if a person can't eat: there's always something. We want these choices. But that doesn't mean we are eager to make the choices ourselves. Instead, most often, we make no choice at all. We fall back on the default, and the default is: Do Something. Fix Something. Is there any way out of this?

There's a school of thought that says the problem is the absence of market forces. If terminal patients—rather than insurance companies or the government—had to pay the added costs for the treatments they chose instead of hospice, they would take the trade-offs into account more. Terminal cancer patients wouldn't pay $80,000 for drugs, and end-stage heart failure patients wouldn't pay $50,000 dollars for defibrillators offering at best a few months extra survival. But this argument ignores an important factor: the people who opt for these treatments aren't thinking a few added months. They're thinking years. They're thinking they're getting at least that lottery ticket's chance that their disease might not even be a problem anymore. Moreover, if there's anything we want to buy in the free market or obtain from our government taxes, it is assurance that, when we find ourselves in need of these options, we won't have to worry about the costs.

This is why the R word—"rationing"—remains such a potent charge. There is broad unease with the circumstances we've found ourselves in but fear of discussing the specifics. For the only seeming alternative to a market solution is outright rationing—death panels, as some have charged. In the 1990s, insurance companies attempted to challenge the treatment decisions of doctors and patients in cases of terminal illness, but the attempts backfired and one case in particular pretty much put an end to strategy—the case of Nelene Fox.

Fox was from Temecula, California, and was diagnosed with metastatic breast cancer in 1991, when she was thirty-eight years old. Surgery and conventional chemotherapy failed, and the cancer spread to her bone marrow. The disease was terminal. Doctors at the University of Southern California offered her a radical but seemingly promising new treatment—high-dose chemotherapy with bone marrow transplantation. To Fox, it was her one chance of cure.

Her insurer, Health Net, denied her request for coverage of the costs, arguing that it was an experimental treatment whose benefits were unproven and that it was therefore excluded under the terms of her policy. The insurer pressed her to get a second opinion from an independent medical center. Fox refused—who were they to tell her to get another opinion? Her life was at stake. Raising $212,000 through charitable donations, she paid the costs of therapy herself, but it was delayed. She died eight months after the treatment. Her husband sued Health Net for bad faith, breach of contract, intentional infliction of emotional damage, and punitive damages and won. The jury awarded her estate $89 million. The HMO executives were branded killers. Ten states enacted laws requiring insurers to pay for bone marrow transplantation for breast cancer.

Never mind that Health Net was right. Research ultimately showed the treatment to have no benefit for breast cancer patients and to actually worsen their lives. But the jury verdict shook the American insurance industry. Raising questions about doctors' and patients' treatment decisions in terminal illness was judged political suicide.

In 2004, executives at another insurance company, Aetna, decided to try a different approach. Instead of reducing aggressive treatment options for their terminally ill policyholders, they decided to try increasing hospice options. Aetna had noted that only a minority of patients ever halted efforts at curative treatment and enrolled in hospice. Even when they did, it was usually not until the very end. So the company decided to experiment: policyholders with a life expectancy of less than a year were allowed to receive hospice services without having to forgo other treatments. A patient like Sara Monopoli could continue to try chemotherapy and radiation and go to the hospital when she wished, but she could also have a hospice team at home focusing on what she needed for the best possible life now and for that morning when she might wake up unable to breathe.

A two-year study of this "concurrent care" program found that enrolled patients were much more likely to use hospice: the figure leaped from 26 percent to 70 percent. That was no surprise, since they weren't forced to give up anything. The surprising result was that they did give up things. They visited the emergency room half as often as the control patients did. Their use of hospitals and ICUs dropped by more than two-thirds. Overall costs fell by almost a quarter.

The result was stunning, and puzzling: it wasn't obvious what made the approach work. Aetna ran a more modest concurrent care program for a broader group of terminally ill patients. For these patients, the traditional hospice rules applied—in order to qualify for home hospice, they had to give up attempts at curative treatment. But either way, they received phone calls from palliative care nurses who offered to check in regularly and help them find services for anything from pain control to making out a living will. For these patients too, hospice enrollment jumped to 70 percent, and their use of hospital services dropped sharply. Among elderly patients, use of intensive care units fell by more than 85 percent. Satisfaction scores went way up. What was going on here? The program's leaders had the impression that they had simply given seriously ill patients someone experienced and knowledgeable to talk to about their daily concerns. Somehow that was enough—just talking.

The explanation would seem to strain credibility, but evidence for it has grown in recent years. Two-thirds of the terminal cancer patients in the Coping with Cancer study reported having had no discussion with their doctors about their goals for end-of-life care, despite being, on average, just four months from death. But the third who did have discussions were far less likely to undergo cardiopulmonary resuscitation or be put on a ventilator or end up in an intensive care unit. Most of them enrolled in hospice. They suffered less, were physically more capable, and were better able, for a longer period, to interact with others. In addition, six months after these patients died, their family members were markedly less likely to experience persistent major depression. In other words, people who had substantive discussions with their doctor about their end-of-life preferences were far more likely to die at peace and in control of their situation and to spare their family anguish.

A landmark 2010 study from the Massachusetts General Hospital had even more startling findings. The researchers randomly assigned 151 patients with stage IV lung cancer, like Sara's, to one of two possible approaches to treatment. Half received usual oncology care. The other half received usual oncology care plus parallel visits with a palliative care specialist. These are specialists in preventing and relieving the suffering of patients, and to see one, no determination of whether they are dying or not is required. If a person has serious, complex illness, palliative specialists are happy to help. The ones in the study discussed with the patients their goals and priorities for if and when their condition worsened. The result: those who saw a palliative care specialist stopped chemotherapy sooner, entered hospice far earlier, experienced less suffering at the end of their lives—*and they lived*

25 percent longer. In other words, our decision making in medicine has failed so spectacularly that we have reached the point of actively inflicting harm on patients rather than confronting the subject of mortality. If end-of-life discussions were an experimental drug, the FDA would approve it.

Patients entering hospice have had no less surprising results. Like many other people, I had believed that hospice care hastens death, because patients forgo hospital treatments and are allowed high-dose narcotics to combat pain. But multiple studies find otherwise. In one, researchers followed 4,493 Medicare patients with either terminal cancer or end-stage congestive heart failure. For the patients with breast cancer, prostate cancer, or colon cancer, the researchers found no difference in survival time between those who went into hospice and those who didn't. And curiously, for some conditions, hospice care seemed to extend survival. Those with pancreatic cancer gained an average of three weeks, those with lung cancer gained six weeks, and those with congestive heart failure gained three months. The lesson seems almost Zen: you live longer only when you stop trying to live longer.

CAN MERE DISCUSSIONS achieve such effects? Consider the case of La Crosse, Wisconsin. Its elderly residents have unusually low end-of-life hospital costs. During their last six months, according to Medicare data, they spend half as many days in the hospital as the national average, and there's no sign that doctors or patients are halting care prematurely. Despite average rates of obesity and smoking, their life expectancy outpaces the national mean by a year.

I spoke to Gregory Thompson, a critical care specialist at Gundersen Lutheran Hospital, while he was on ICU duty one evening, and he ran through his list of patients with me. In most respects, the patients were like those found in any ICU—terribly sick and living through the most perilous days of their lives. There was a young woman with multiple organ failure from a devastating case of pneumonia, a man in his midsixties with a ruptured colon that had caused a rampaging infection and a heart attack. Yet these patients were completely different from those in the ICUs I'd worked in: none had a terminal disease; none battled the final stages of metastatic cancer or untreatable heart failure or dementia.

To understand La Crosse, Thompson said, you had to go back to 1991, when local medical leaders headed a systematic campaign to get medical people and patients to discuss end-of-life wishes. Within a few years, it became routine for all patients admitted to a hospital, nursing home, or assisted living facility to sit down with someone experienced in these conversations and complete a multiple-choice form that boiled down to four crucial questions. At this moment in your life, the form asked:

1. Do you want to be resuscitated if your heart stops?
2. Do you want aggressive treatments such as intubation and mechanical ventilation?
3. Do you want antibiotics?
4. Do you want tube or intravenous feeding if you can't eat on your own?

By 1996, 85 percent of La Crosse residents who died had a written advanced directive like this, up from 15 percent, and doctors virtually always knew of the instructions and followed them. Having this system in place, Thompson said, has made his job vastly easier. But it's not because the specifics are spelled out for him every time a sick patient arrives in his unit.

"These things are not laid out in stone," he told me. Whatever the yes/no answers people may put on a piece of paper, one will find nuances and complexities in what they mean. "But instead of having the discussion when they get to the ICU, we find many times it has already taken place."

Answers to the list of questions change as patients go from entering the hospital for the delivery of a child to entering for complications of Alzheimer's disease. But in La Crosse, the system means that people are far more likely to have talked about what they want and what they don't want before they and their relatives find themselves in the throes of crisis and fear. When wishes aren't clear, Thompson said, "families have also become much more receptive to having the discussion." The discussion, not the list, was what mattered most. Discussion had brought La Crosse's end-of-life costs down to half the national average. It was that simple—and that complicated.

ONE WINTER SATURDAY morning, I met with a woman I had operated on the night before. She had been undergoing a procedure for the removal of an ovarian cyst when the gynecologist who was operating on her discovered that she had metastatic colon cancer. I was summoned, as a general surgeon, to see what could be done. I removed a section of her colon that had a large cancerous mass, but the cancer had already spread widely. I had not been able to get it all. Now I introduced myself. She said a resident had told her that a tumor was found and part of her colon had been excised.

Yes, I said. I'd been able to take out "the main area of involvement." I explained how much bowel was removed, what the recovery would be like—everything except how much cancer there was.

But then I remembered how timid I'd been with Sara Monopoli, and all those studies about how much doctors beat around the bush. So when she asked me to tell her more about the cancer, I explained that it had spread not only to her ovaries but also to her lymph nodes. I said that it had not been possible to remove all the disease. But I found myself almost immediately minimizing what I'd said. "We'll bring in an oncologist," I hastened to add. "Chemotherapy can be very effective in these situations."

She absorbed the news in silence, looking down at the blankets drawn over her mutinous body. Then she looked up at me. "Am I going to die?"

I flinched. "No, no," I said. "Of course not."

A few days later, I tried again. "We don't have a cure," I explained. "But treatment can hold the disease down for a long time." The goal, I said, was to "prolong your life" as much as possible.

I have followed her in the months and years since, as she embarked on chemotherapy. She has done well. So far, the cancer is in check. Once, I asked her and her husband about our initial conversations. They didn't remember them very fondly. "That one phrase that you used—'prolong your life'—it just . . ." She didn't want to sound critical.

"It was kind of blunt," her husband said.

"It sounded harsh," she echoed. She felt as if I'd dropped her off a cliff.

I spoke to Susan Block, a palliative care specialist at my hospital who has had thousands of these difficult conversations and is a nationally recognized pioneer in training doctors and others in managing end-of-life issues with patients and their families. "You have to understand," Block told me. "A family meeting is a procedure, and it requires no less skill than performing an operation."

One basic mistake is conceptual. To most doctors, the primary purpose of a discussion about terminal illness is to determine what people want—whether they want chemo or not, whether they want to be resuscitated or not, whether they want hospice or not. We focus on laying out the facts and the options. But that's a mistake, Block said.

"A large part of the task is helping people negotiate the overwhelming anxiety—anxiety about death, anxiety about suffering, anxiety about loved ones, anxiety about finances," she explained. "There are many worries and real terrors." No one conversation can address them all. Arriving at an acceptance of one's mortality and a clear understanding of the limits and the possibilities of medicine is a process, not an epiphany.

There is no single way to take people with terminal illness through the process, but there are some rules, according to Block. You sit down. You make time. You're not determining whether they want treatment X versus Y. You're trying to learn what's most important to them under the circumstances—so that you can provide information and advice on the approach that gives them their best chance of achieving it. This process requires as much listening as talking. If you are talking more than half of the time, Block says, you're talking too much.

The words you use matter. According to palliative specialists, you shouldn't say, "I'm sorry things turned out this way," for example. It can sound like you're distancing yourself. You should say, "I wish things were different." You don't ask, "What do you want when you are dying?" You ask, "If time becomes short, what is most important to you?"

Block has a list of questions that she aims to cover with sick patients in the time before decisions have to be made: What do they understand their prognosis to be, what are their concerns about what lies ahead, what kinds of trade-offs are they willing to make, how do they want to spend their time if their health worsens, who do they want to make decisions if they can't?

A decade earlier, her seventy-four-year-old father, Jack Block, a professor emeritus of psychology at the University of California at Berkeley, was admitted to a San Francisco hospital with symptoms from what proved to be a mass growing in the spinal cord of his neck. She flew out to see him. The neurosurgeon said that the procedure to remove the mass carried a 20 percent chance of leaving him quadriplegic, paralyzed from the neck down. But without it he had a 100 percent chance of becoming quadriplegic.

The evening before surgery, father and daughter chatted about friends and family, trying to keep their minds off what was to come, and then she left for the night. Halfway across the Bay Bridge, she recalled, "I realized, 'Oh, my God, I don't know what he really wants.'" He'd made her his health care proxy, but they had talked about such situations only superficially. So she turned the car around.

Going back in "was really uncomfortable," she said. It made no difference that she was an expert in end-of-life discussions. "I just felt awful having the conversation with my dad." But she went through her list. She told him, "'I need to understand how much you're willing to go through to have a shot at being alive and what level of being alive is tolerable to you.' We had this quite agonizing conversation where he said—and this totally shocked me—'Well, if I'm able to eat chocolate ice cream and watch football on TV, then I'm willing to stay alive. I'm willing to go through a lot of pain if I have a shot at that.'"

"I would never have expected him to say that," Block said. "I mean, he's a professor emeritus. He's never watched a football game in my conscious memory. The whole picture—it wasn't the guy I thought I knew." But the conversation proved critical, because after surgery he developed bleeding in the spinal cord. The surgeons told her that in order to save his life they would need to go back in. But the bleeding had already made him nearly quadriplegic, and he would remain severely disabled for many months and likely forever. What did she want to do?

"I had three minutes to make this decision, and I realized, he had already made the decision." She asked the surgeons whether, if her father survived, he would still be able to eat chocolate ice cream and watch football on TV. Yes, they said. She gave the okay to take him back to the operating room.

"If I had not had that conversation with him," she told me, "my instinct would have been to let him go at that moment because it just seemed so awful. And I would have beaten myself up. Did I let him go too soon?" Or she might have gone ahead and sent him to surgery, only to find—as occurred—that he was faced with a year of "very horrible rehab" and disability. "I would have felt so guilty that I condemned him to that," she said. "But there was no decision for me to make." He had decided.

During the next two years, he regained the ability to walk short distances. He required caregivers to bathe and dress him. He had difficulty swallowing and eating. But his mind was intact and he had partial use of his hands—enough to write two books and more than a dozen scientific articles. He lived for ten years after the operation. Eventually, however, his difficulties with swallowing advanced to the point where he could not eat without aspirating food particles, and he cycled between hospital and rehabilitation facilities with the pneumonias that resulted. He didn't want a feeding tube. And it became evident that the battle for the dwindling chance of a miraculous recovery was going to leave him unable ever to go home again. So, just a few months before I'd spoken with Block, her father decided to stop the battle and go home.

"We started him on hospice care," Block said. "We treated his choking and kept him comfortable. Eventually, he stopped eating and drinking. He died about five days later."

SUSAN BLOCK AND her father had the conversation that we all need to have when the chemotherapy stops working, when we start needing oxygen at home, when we face high-risk surgery, when the liver failure keeps progressing, when we become unable to dress ourselves. I've heard Swedish doctors call it a "breakpoint discussion," a series of conversations to sort out when they need to switch from fighting for time to fighting for the other things that people value—being with family or traveling or enjoying chocolate ice cream. Few people have these conversations, and there is good reason for anyone to dread them. They can unleash difficult emotions. People can become angry or overwhelmed. Handled poorly, the conversations can cost a person's trust. Handled well, they can take real time.

I spoke to an oncologist who told me about a twenty-nine-year-old patient she had recently cared for who had an inoperable brain tumor that continued to grow through second-line chemotherapy. The patient elected not to attempt any further chemotherapy, but getting to that decision required hours of discussion, for this was not the decision he had expected to make. First, the oncologist said, she had a discussion with him alone. They reviewed the story of how far he'd come, the options that remained. She was frank. She told him that in her entire career she had never seen third-line chemotherapy produce a significant response in his type of brain tumor. She had looked for experimental therapies, and none were truly promising. And, although she was willing to proceed with chemotherapy, she told him how much strength and time the treatment would take away from him and his family.

He did not shut down or rebel. His questions went on for an hour. He asked about this therapy and that therapy. Gradually, he began to ask about what would happen as the tumor got bigger, what symptoms he'd have, what ways they could try to control them, how the end might come.

The oncologist next met with the young man together with his family. That discussion didn't go so well. He had a wife and small children, and at first his wife wasn't ready to contemplate stopping chemo. But when the oncologist asked the patient to explain in his own words what they'd discussed, she understood. It was the same with his mother, who was a nurse. Meanwhile, his father sat quietly and said nothing the entire time.

A few days later, the patient returned to talk to the oncologist. "There should be something. There must be something," he said. His father had shown him reports of cures on the Internet. He confided how badly his father was taking the news. No patient wants to cause his family pain. According to Block, about two-thirds of patients are willing to undergo therapies they don't want if that is what their loved ones want.

The oncologist went to the father's home to meet with him. He had a sheaf of possible trials and treatments printed from the Internet. She went through them all. She was willing to change her opinion, she told him. But either the treatments were for brain tumors that were very different from his son's or else he didn't qualify. None were going to be miraculous. She told the father that he needed to understand: time with his son was limited, and the young man was going to need his father's help getting through it.

The oncologist noted wryly how much easier it would have

been for her just to prescribe the chemotherapy. "But that meeting with the father was the turning point," she said. The patient and the family opted for hospice. They had more than a month together before he died. Later, the father thanked the doctor. That last month, he said, the family simply focused on being together, and it proved to be the most meaningful time they'd ever spent.

Given how prolonged some of these conversations have to be, many people argue that the key problem has been the financial incentives: we pay doctors to give chemotherapy and to do surgery but not to take the time required to sort out when to do so is unwise. This certainly is a factor. But the issue isn't merely a matter of financing. It arises from a still unresolved argument about what the function of medicine really is—what, in other words, we should and should not be paying for doctors to do.

The simple view is that medicine exists to fight death and disease, and that is, of course, its most basic task. Death is the enemy. But the enemy has superior forces. Eventually, it wins. And in a war that you cannot win, you don't want a general who fights to the point of total annihilation. You don't want Custer. You want Robert E. Lee, someone who knows how to fight for territory that can be won and how to surrender it when it can't, someone who understands that the damage is greatest if all you do is battle to the bitter end.

More often, these days, medicine seems to supply neither Custers nor Lees. We are increasingly the generals who march the soldiers onward, saying all the while, "You let me know when you want to stop." All-out treatment, we tell the incurably ill, is a train you can get off at any time—just say when. But for most patients and their families we are asking too much. They remain riven by doubt and fear and desperation; some are deluded by a fantasy of what medical science can achieve. Our responsibility, in medicine, is to deal with human beings as they are. People die only once. They have no experience to draw on. They need doctors and nurses who are willing to have the hard discussions and say what they have seen, who will help people prepare for what is to come—and escape a warehoused oblivion that few really want.

SARA MONOPOLI HAD had enough discussions to let her family and her oncologist know that she did not want hospitals or ICUs at the end—but not enough to have learned how to achieve her goal. From the moment she arrived in the emergency room that Friday morning in February, the train of events ran against a peaceful ending. There was one person who was disturbed by this, though, and who finally decided to intercede—Chuck Morris, her primary care physician. As her illness had progressed through the previous year, he had left the decision making largely to Sara, her family, and the oncology team. Still, he had seen her and her husband regularly and listened to their concerns. That desperate morning, Morris was the one person Rich called before getting into the ambulance. He headed to the emergency room and met Sara and Rich when they arrived.

Morris said that the pneumonia might be treatable. But he told Rich, "I'm worried this is it. I'm really worried about her." And he told him to let the family know that he said so.

Upstairs in her hospital room, Morris talked with Sara and Rich about the ways in which the cancer had been weakening her, making it hard for her body to fight off infection. Even if the antibiotics halted the infection, he said, he wanted them to remember that there was nothing that would stop the cancer.

Sara looked ghastly, Morris told me. "She was so short of

breath. It was uncomfortable to watch. I still remember the attending"—the covering oncologist who admitted her for the pneumonia treatment. "He was actually kind of rattled about the whole case, and for him to be rattled is saying something."

After her parents arrived, Morris talked with them too, and when they were finished Sara and her family agreed on a plan. The medical team would continue the antibiotics. But if things got worse, they would not put her on a breathing machine. They also let him call the palliative care team to visit. The team prescribed a small dose of morphine, which immediately eased her breathing. Her family saw how much her suffering diminished, and suddenly they didn't want any more suffering. The next morning, they were the ones to hold back the medical team.

"They wanted to put a catheter in her, do this other stuff to her," her mother, Dawn, told me. "I said, 'No. You aren't going to do anything to her.' I didn't care if she wet her bed. They wanted to do lab tests, blood pressure measurements, finger sticks. I was very uninterested in their bookkeeping. I went over to see the head nurse and told them to stop."

In the previous three months, almost nothing we'd done to Sara—none of the scans or tests or radiation or extra rounds of chemotherapy—had likely achieved anything except to make her worse. She may well have lived longer without any of it. At least she was spared at the very end.

That day, Sara fell into unconsciousness as her body continued to fail. Through the next night, Rich recalled, "there was this awful groaning." There is no prettifying death. "Whether it was with inhaling or exhaling, I don't remember, but it was horrible, horrible, horrible to listen to."

Her father and her sister still thought that she might rally. But when the others had stepped out of the room, Rich knelt down weeping beside Sara and whispered in her ear. "It's okay to let go," he said. "You don't have to fight anymore. I will see you soon."

Later that morning, her breathing changed, slowing. Rich said, "Sara just kind of startled. She let a long breath out. Then she just stopped."

7 · *Hard Conversations*

Traveling abroad sometime afterward, I fell into a conversation with two doctors from Uganda and a writer from South Africa. I told them about Sara's case and asked what they thought should have been done for her. To their eyes, the choices we offered her seemed extravagant. Most people with terminal illness in their countries would never have come to the hospital, they said. Those who did would neither expect nor tolerate the extremes of multiple chemotherapy regimens, last-ditch surgical procedures, experimental therapies—when the problem's ultimate outcome was so dismally clear. And the health system wouldn't have the money for it.

But then they couldn't help but talk about their own experiences, and their tales sounded familiar: a grandparent put on life support against his wishes, a relative with incurable liver cancer who died in the hospital on an experimental treatment, a brother-in-law with a terminal brain tumor who nonetheless endured endless cycles of chemotherapy that had no effect except to cut him down further and further. "Each round was more horrible than the last," the South African writer told me. "I saw the medicine eat his flesh. The children are still traumatized. He could never let go."

Their countries were changing. Five of the ten fastest-growing economies in the world are in Africa. By 2030, one-half to two-thirds of the global population will be middle class. Vast numbers of people are becoming able to afford consumer goods like televisions and cars—and health care. Surveys in some African cities are finding, for example, that half of the elderly over eighty years old now die in the hospital and even higher percentages of those less than eighty years old do. These are numbers that actually exceed those in most developed countries today. Versions of Sara's story are becoming global. As incomes rise, private sector health care is increasing rapidly, usually paid for in cash. Doctors everywhere become all too ready to offer false hopes, leading families to empty bank accounts, sell their seed crops, and take money from their children's education for futile treatments. Yet at the same time, hospice programs are appearing everywhere from Kampala to Kinshasa, Lagos to Lesotho, not to mention Mumbai to Manila.

Scholars have posited three stages of medical development that countries go through, paralleling their economic development. In the first stage, when a country is in extreme poverty, most deaths occur in the home because people don't have access to professional diagnosis and treatment. In the second stage, when a country's economy develops and its people transition to higher income levels, the greater resources make medical capabilities more widely available. People turn to health care systems when they are ill. At the end of life, they often die in the hospital instead of the home. In the third stage, as a country's income climbs to the highest levels, people have the means to become concerned about the quality of their lives, even in sickness, and deaths at home actually rise again.

This pattern seems to be what is happening in the United States. Whereas deaths in the home went from a clear majority in 1945 to just 17 percent in the late eighties, since the nineties the numbers have reversed direction. Use of hospice care has been growing steadily—to the point that, by 2010, 45 percent of Americans died in hospice. More than half of them received hospice care at home, and the remainder received it in an institution, usually an inpatient hospice facility for the dying or a nursing home. These are among the highest rates in the world.

A monumental transformation is occurring. In this country and across the globe, people increasingly have an alternative to withering in old age homes and dying in hospitals—and millions of them are seizing the opportunity. But this is an unsettled time. We've begun rejecting the institutionalized version of aging and death, but we've not yet established our new norm. We're caught in a transitional phase. However miserable the old system has been, we are all experts at it. We know the dance moves. You agree to become a patient, and I, the clinician, agree to try to fix you, whatever the improbability, the misery, the damage, or the cost. With this new way, in which we together try to figure out how to face mortality and preserve the fiber of a meaningful life, with its loyalties and individuality, we are plodding novices. We are going through a societal learning curve, one person at a time. And that would include me, whether as a doctor or as simply a human being.

MY FATHER WAS in his early seventies when I was forced to realize that he might not be immortal. He'd been as healthy as a Brahma bull, playing tennis three days a week, maintaining a busy urology practice, and serving as president of the local Rotary Club. He had tremendous energy. He did numerous charity projects, including working with a rural Indian college he'd established, expanding it from a single building to a campus with some two thousand students. Whenever I came home, I'd bring my tennis rackets and we'd go out on the local courts. He played to win, and so did I. He'd drop shot me; I'd drop shot him. He'd lob me; I'd lob him. He had picked up a few old-man habits, like blowing his nose onto the court whenever he felt like it or making me chase down our errant tennis balls. But I took them to be the kinds of advantages a father takes with a son, rather than signs of age. In more than thirty years of medical practice, he'd not canceled his clinic or operating schedule for sickness once. So when he mentioned the development of a neck pain that shot down his left arm and caused tingling in the tips of his left fingers, neither one of us was inclined to think too much of it. An X-ray of his neck showed only arthritis. He took anti-inflammatory medication, underwent physical therapy, and took a break from using an overhead serve, which exacerbated the pain. Otherwise it was life as usual for him.

Over the next couple years, however, the neck pain progressed. It became difficult for him to sleep comfortably. The tingling in the tips of his left fingers became full-blown numbness and spread to his whole left hand. He found he had trouble feeling the thread when tying sutures during vasectomies. In the spring of 2006, his doctor ordered an MRI of his neck. The findings were a complete shock. The scan revealed a tumor growing inside his spinal cord.

That was the moment when we stepped through the looking glass. Nothing about my father's life and expectations for it would remain the same. Our family was embarking on its own confrontation with the reality of mortality. The test for us as parents and children would be whether we could make the path go any differently for my dad than I, as a doctor, had made it go for my patients. The No. 2 pencils had been handed out. The timer had been started. But we had not even registered that the test had begun.

My father sent me the images by e-mail, and we spoke by phone as we looked at them on our laptops. The mass was nauseating to behold. It filled the entire spinal canal, extending all the way up to the base of his brain and down to the level of his shoulder blades. It appeared to be obliterating his spinal cord. I was amazed that he wasn't paralyzed, that all the thing had done so far was make his hand numb and his neck hurt. We didn't talk about any of this, though. We had trouble finding anywhere safe for conversation to take purchase. I asked him what the radiologist's report said the mass might be. Various benign and malignant tumors were listed, he said. Did it suggest any other possibilities besides a tumor? Not really, he said. Two surgeons, we puzzled over how a tumor like this could be removed. But there seemed no way, and we grew silent. Let's talk to a neurosurgeon before jumping to any conclusions, I said.

Spinal cord tumors are rare, and few neurosurgeons have much experience with them. A dozen cases is a lot. Among the most experienced neurosurgeons was one at the Cleveland Clinic, which was two hundred miles from my parents' home, and one at my hospital in Boston. We made appointments at both places.

Both surgeons offered surgery. They would open up the spinal cord—I didn't even know that was possible—and remove as much of the tumor as they could. They'd only be able to remove part of it, though. The tumor's primary source of damage was from its growth inside the confined space of the spinal canal—the beast was outgrowing its cage. The expansion of the mass was crushing the spinal cord against the vertebral bone, causing pain as well as destruction of the nerve fibers that make up the cord. So both surgeons proposed also doing a procedure to expand the space for the tumor to grow. They'd decompress the tumor, by opening the back of the spinal column, and stabilize the vertebrae with rods. It'd be like taking the back wall off a tall building and replacing it with columns to hold up the floors.

The neurosurgeon at my hospital advocated operating right away. The situation was dangerous, he told my father. He could become quadriplegic in weeks. No other options existed—chemotherapy and radiation were not nearly as effective in stopping progression as surgery. The operation had risks, he said, but he wasn't too worried about them. He was more concerned about the tumor. My father needed to act before it was too late.

The neurosurgeon at the Cleveland Clinic painted a more ambiguous picture. While he offered the same operation, he didn't push to do it right away. He said that while some spinal cord tumors advance rapidly, he'd seen many take years to progress, and they did so in stages, not all at once. He didn't think my father would go from a numb hand to total paralysis overnight. The question therefore was when to go in, and he believed that should be when the situation became intolerable enough for my father to want to attempt treatment. The surgeon was not as blithe about its risks as the other neurosurgeon. He thought it carried a one in four chance of itself causing quadriplegia or death. My father, he said, would "need to draw a line in the sand." Were his symptoms already bad enough that he wanted surgery now? Would he want to wait until he started to feel hand symptoms that threatened his ability to do surgery? Would he want to wait until he couldn't walk?

The information was difficult to take in. How many times had my father given patients bad news like this—that they had prostate cancer, for instance, requiring similarly awful choices to be made. How many times had I done the same? The news, nonetheless, came like a body blow. Neither surgeon came out and said that the tumor was fatal, but neither said the tumor could be removed, either. It could only be "decompressed."

In theory, a person should make decisions about life and death matters analytically, on the basis of the facts. But the facts were shot through with holes and uncertainties. The tumor was rare. No clear predictions could be made. Making choices required somehow filling the gaps, and what my father filled them with was fear. He feared the tumor and what it would do to him, and he also feared the solution being proposed. He could not fathom opening up the spinal cord. And he found it difficult to put his trust in any operation that he did not understand—that he did not feel capable of doing himself. He asked the surgeons numerous questions about how exactly it would be done. What kind of instrument do you use to enter the spinal cord, he asked? Do you use a microscope? How do you cut through the tumor? How do you cauterize the blood vessels? Couldn't the cautery damage the nerve fibers of the cord? We use such and such an instrument to control prostate bleeding in urology—wouldn't it be better to use that? Why not?

The neurosurgeon at my hospital didn't much like my father's questions. He was fine answering the first couple. But after that he grew exasperated. He had the air of the renowned professor he was—authoritative, self-certain, and busy with things to do.

Look, he said to my father, the tumor was dangerous. He, the neurosurgeon, had a lot of experience treating such tumors. Indeed, no one had more. The decision for my father was whether he wanted to do something about his tumor. If he did, the neurosurgeon was willing to help. If he didn't, that was his choice.

When the doctor finished, my father didn't ask any more questions. But he'd also decided that this man wasn't going to be his surgeon.

The Cleveland Clinic neurosurgeon, Edward Benzel, exuded no less confidence. But he recognized that my father's questions came from fear. So he took the time to answer them, even the annoying ones. Along the way, he probed my father, too. He said that it sounded like he was more worried about what the operation might do to him than what the tumor would.

My father said he was right. My father didn't want to risk losing his ability to practice surgery for the sake of treatment of uncertain benefit. The surgeon said that he might feel the same way himself in my father's shoes.

Benzel had a way of looking at people that let them know he was really looking at them. He was several inches taller than my parents, but he made sure to sit at eye level. He turned his seat away from the computer and planted himself directly in front of them. He did not twitch or fidget or even react when my father talked. He had that midwesterner's habit of waiting a beat after people have spoken before speaking himself, in order to see if they are really done. He had small, dark eyes set behind wire-rim glasses and a mouth hidden by the thick gray bristle of a Van Dyke beard. The only thing to hint at what he was thinking was the wrinkle of his glossy dome of a forehead. Eventually, he steered the conversation back to the central issue. The tumor was worrisome, but he now understood something about my father's concerns. He believed my father had time to wait and see how quickly his symptoms changed. He could hold off surgery until he felt he needed it. My father decided to go with Benzel and his counsel. My parents made a plan to return in a few months for a checkup and to call sooner if he experienced any signs of serious change.

Did he prefer Benzel simply because he'd portrayed a better, at least slightly less alarming picture of what might happen with the tumor? Maybe. It happens. Patients tend to be optimists, even if that makes them prefer doctors who are more likely to be wrong. Only time would tell which of the two surgeons was right. Nonetheless, Benzel had made the effort to understand what my father cared about most, and to my father that counted for a lot. Even before the visit was halfway over, he had decided Benzel was the one he would trust.

In the end, Benzel was also the one who proved right. As time passed, my father noticed no change in symptoms. He decided to put off the follow-up appointment. It was ultimately a year before he returned to see Benzel. A repeat MRI showed the tumor had enlarged. Yet physical examination found no diminishment in my dad's strength, sensation, or mobility. So they decided to go primarily by how he felt, not by what the pictures looked like. The MRI reports would say haunting things, like the imaging "demonstrates significant increase in size of the cervical mass at the level of the medulla and midbrain." But for months at a stretch, nothing occurred to change anything relevant for how he lived.

The neck pain remained annoying, but my father figured out the best positions for sleeping at night. When chilly weather came, he found that his numb left hand became bone-cold. He took to wearing a glove over it, Michael Jackson–style, even indoors at home. Otherwise, he kept on driving, playing tennis, doing surgery, living life as he had been. He and his neurosurgeon knew what was coming. But they also knew what mattered to him and left well enough alone. This was, I remember thinking, just the way I ought to make decisions with my own patients—the way we all ought to in medicine.

DURING MEDICAL SCHOOL, my fellow classmates and I were assigned to read a short paper by two medical ethicists, Ezekiel and Linda Emanuel, on the different kinds of relationships that we, as budding new clinicians, might have with our patients. The oldest, most traditional kind is a paternalistic relationship—we are medical authorities aiming to ensure that patients receive what we believe best for them. We have the knowledge and experience. We make the critical choices. If there were a red pill and a blue pill, we would tell you, "Take the red pill. It will be good for you." We might tell you about the blue pill; but then again, we might not. We tell you only what we believe you need to know. It is the priestly, doctor-knows-best model, and although often denounced it remains a common mode, especially with vulnerable patients—the frail, the poor, the elderly, and anyone else who tends to do what they're told.

The second type of relationship the authors termed "informative." It's the opposite of the paternalistic relationship. We tell you the facts and figures. The rest is up to you. "Here's what the red pill does, and here's what the blue pill does," we would say. "Which one do you want?" It's a retail relationship. The doctor is the technical expert. The patient is the consumer. The job of doctors is to supply up-to-date knowledge and skills. The job of patients is to supply the decisions. This is the increasingly common way for doctors to be, and it tends to drive us to become ever more specialized. We know less and less about our patients but more and more about our science. Overall, this kind of relationship can work beautifully, especially when the choices are clear, the trade-offs are straightforward, and people have clear preferences. You get only the tests, the pills, the operations, the risks that you want and accept. You have complete autonomy.

The neurosurgeon at my hospital in Boston showed elements of both these types of roles. He was the paternalistic doctor: surgery was my father's best choice, he insisted, and my father needed to have it now. But my father pushed him to try to be the informative doctor, to go over the details and the options. So the surgeon switched, but the descriptions only increased my father's fears, fueled more questions, and made him even more uncertain about what he preferred. The surgeon didn't know what to do with him.

In truth, neither type is quite what people desire. We want information and control, but we also want guidance. The Emanuels described a third type of doctor-patient relationship, which they called "interpretive." Here the doctor's role is to help patients determine what they want. Interpretive doctors ask, "What is most important to you? What are your worries?" Then, when they know your answers, they tell you about the red pill and the blue pill and which one would most help you achieve your priorities.

Experts have come to call this shared decision making. It seemed to us medical students a nice way to work with patients as physicians. But it seemed almost entirely theoretical. Certainly, to the larger medical community, the idea that most doctors would play this kind of role for patients seemed far-fetched at the time. (Surgeons? "Interpretive?" Ha!) I didn't hear clinicians talk about the idea again and largely forgot about it. The choices in training seemed to be between the more paternalistic style and the more informative one. Yet, less than two decades later, here we were with my father, in a neurosurgeon's office in Cleveland, Ohio, talking about MRI images showing a giant and deadly tumor growing in his spinal cord, and this other kind of doctor— one willing to genuinely share decision making—was precisely what we found. Benzel saw himself as neither the commander nor a mere technician in this battle but instead as a kind of counselor and contractor on my father's behalf. It was exactly what my father needed.

Rereading the paper afterward, I found the authors warning

that doctors would sometimes have to go farther than just interpreting people's wishes in order to serve their needs adequately. Wants are fickle. And everyone has what philosophers call "second-order desires"—desires about our desires. We may wish, for instance, to be less impulsive, more healthy, less controlled by primitive desires like fear or hunger, more faithful to larger goals. Doctors who listen to only the momentary, first-order desires may not be serving their patients' real wishes, after all. We often appreciate clinicians who push us when we make shortsighted choices, such as skipping our medications or not getting enough exercise. And we often adjust to changes we initially fear. At some point, therefore, it becomes not only right but also necessary for a doctor to deliberate with people on their larger goals, to even challenge them to rethink ill-considered priorities and beliefs.

In my career, I have always been most comfortable being Dr. Informative. (My generation of physicians has mostly steered away from being Dr. Knows-Best.) But Dr. Informative was clearly not sufficient to help Sara Monopoli or the many other seriously ill patients I'd had.

Around the time of my father's visits with Benzel, I was asked to see a seventy-two-year-old woman with metastatic ovarian cancer who had come to my hospital's emergency room because of vomiting. Her name was Jewel Douglass, and looking through her medical records, I saw that she'd been in treatment for two years. Her first sign of the cancer had been a feeling of abdominal bloating. She saw her gynecologist, who found, with the aid of an ultrasound, a mass in her pelvis the size of a child's fist. In the operating room, it proved to be an ovarian cancer, and it had spread throughout her abdomen. Soft, fungating tumor deposits studded her uterus, her bladder, her colon, and the lining of her abdomen. The surgeon removed both of her ovaries, the whole of her uterus, half of her colon, and a third of her bladder. She underwent three months of chemotherapy. With this kind of treatment, most ovarian cancer patients at her stage survive two years and a third survive five years. About 20 percent of patients are actually cured. She hoped to be among these few.

She reportedly tolerated the chemotherapy well. She'd lost her hair but otherwise experienced only mild fatigue. At nine months, no tumor could be seen on her CT scans at all. At one year, however, a scan showed a few pebbles of tumor had grown back. She felt nothing—they were just millimeters in size—but there they were. Her oncologist started a different chemotherapy regimen. This time Douglass had more painful side effects—mouth sores, a burn-like rash across her body—but with salves of various kinds they were tolerable. A follow-up scan showed the treatment hadn't worked, though. The tumors grew. They began giving her shooting pains in her pelvis.

She switched to a third kind of chemotherapy. This one was more effective—the tumors shrank, the shooting pains went away—but the side effects were much worse. Her records reported her having terrible nausea despite trying multiple medications to stop it. Limb-sapping fatigue put her in bed for hours a day. An allergic reaction gave her hives and intense itching that required steroid pills to control. One day, she became severely short of breath and had to be brought to the hospital by ambulance. Tests showed she had developed pulmonary emboli, just as Sara Monopoli had. She was put on daily injections of a blood thinner and only gradually regained her ability to breathe normally.

Then she developed clenching, gas-like pains in her belly. She began vomiting. She found she could not hold anything down, liquid or solid. She called her oncologist, who ordered a CT scan. It showed a blockage in a loop of her bowel caused by her metastases. She was sent from the radiology department to the emergency room. As the general surgeon on duty, I was called to see what I could do.

I reviewed the images from her scan with a radiologist, but we could not precisely make out how the cancer was causing her intestinal blockage. It was possible that the bowel loop had gotten caught on a knuckle of tumor and then twisted—a problem that could potentially resolve on its own, if given time. Or else the bowel had become physically compressed by a tumor growth—a problem that would resolve only with surgery to either remove or bypass the obstruction. Either way, it was a troubling sign of the advancement of her cancer—despite, now, three regimens of chemotherapy.

I went to talk to Douglass, thinking about exactly how much of this to confront her with. By this time, a nurse had given her intravenous fluids and a resident had inserted a three-foot-long tube into her nose down to her stomach, which had already drained out a half liter of bile-green fluid. Nasogastric tubes are uncomfortable, torturous devices. People who have the things stuck into them are usually not in a conversational mood. When I introduced myself, however, she smiled, made a point of having me repeat my name, and made sure she could pronounce it correctly. Her husband sat by her in a chair, pensive and quiet, letting her take the lead.

"I seem to be in a pickle from what I understand," she said.

She was the sort of person who'd managed, even with the tube taped into her nose, to fix her hair, which she wore in a bob, put her glasses back on, and smooth her hospital sheets over herself neatly. She was doing her best to maintain her dignity under the circumstances.

I asked how she was feeling. The tube had helped, she said. She felt much less nauseated.

I asked her to explain what she'd been told. She said, "Well, doctor, it seems my cancer is blocking me up. So everything that goes down comes back up again."

She'd grasped the grim basics perfectly. At this point, we had no especially difficult decisions to make. I told her there was a chance that this was just a twist in a bowel loop and that with a day or two's time it might open up on its own. If it didn't, I said, we'd have to talk about possibilities like surgery. Right now, though, we could wait.

I was not yet willing to raise the harder issue. I could have pushed ahead, trying to be hard-nosed, and told her that, no matter what happened, this blockage was a bad harbinger. Cancers kill people in many ways, and gradually taking away their ability to eat is one of them. But she didn't know me, and I didn't know her. I decided I needed time before attempting that line of discussion.

A day later, the news was as good as could be hoped. First, the fluid flowing out of the tube slowed down. Then she started passing gas and having bowel movements. We were able to remove her nasogastric tube and feed her a soft, low-roughage diet. It looked like she would be fine for now.

I was tempted simply to discharge her home and wish her well—to skip the hard conversation altogether. But this wasn't likely to be the end of the matter for Douglass. So before she left, I returned to her hospital room and sat with her, her husband, and one of her sons.

I started out saying how pleased I was to see her eating again. She said she'd never been so happy to pass gas in her life. She had questions about the foods she should eat and the ones she shouldn't in order to avoid blocking up her bowel again, and

I answered them. We made some small talk, and her family told me a bit about her. She'd once been a singer. She became Miss Massachusetts 1956. Afterward, Nat King Cole asked her to join his tour as a backup singer. But she discovered that the life of an entertainer was not what she wanted. So she came home to Boston. She met Arthur Douglass, who took over his family's funeral home business after they married. They raised four children but suffered through the death of their oldest child, a son, at a young age. She was looking forward to getting home to her friends and family and to taking a trip to Florida they had planned to get away from all this cancer business. She was eager to leave the hospital.

Nonetheless, I decided to push. Here was an opening to discuss her future, and I realized it was one I needed to take. But how to do it? Was I just to blurt out, "By the way, the cancer is getting worse and will probably block you up, again"? Bob Arnold, a palliative care physician I'd met from the University of Pittsburgh, had explained to me that the mistake clinicians make in these situations is that they see their task as just supplying cognitive information—hard, cold facts and descriptions. They want to be Dr. Informative. But it's the meaning behind the information that people are looking for more than the facts. The best way to convey meaning is to tell people what the information means to you yourself, he said. And he gave me three words to use to do that.

"I am worried," I told Douglass. The tumor was still there, I explained, and I was worried the blockage was likely to come back.

They were such simple words, but it wasn't hard to sense how much they communicated. I had given her the facts. But by including the fact that I was worried, I'd not only told her about the seriousness of the situation, I'd told her that I was on her side—I was pulling for her. The words also told her that, although I feared something serious, there remained uncertainties—possibilities for hope within the parameters nature had imposed.

I let her and her family take in what I'd said. I don't remember Douglass's precise words when she spoke, but I remember that the weather in the room had changed. Clouds rolled in. She wanted more information. I asked her what she wanted to know.

This was another practiced and deliberate question on my part. I felt foolish to still be learning how to talk to people at this stage of my career. But Arnold had also recommended a strategy palliative care physicians use when they have to talk about bad news with people—they "ask, tell, ask." They ask what you want to hear, then they tell you, and then they ask what you understood. So I asked.

Douglass said she wanted to know what could happen to her. I said that it was possible that nothing like this episode would ever happen again. I was concerned, however, that the tumor would likely cause another blockage. She'd have to return to the hospital in that case. We'd have to put the tube back in. Or I might need to do surgery to relieve the blockage. That could require giving her an ileostomy, a rerouting of her small bowel to the surface of her skin where we would attach the opening to a bag. Or I might not be able to relieve the blockage at all.

She didn't ask any more questions after that. I asked her what she'd understood. She said she understood that she wasn't out of trouble. And with those words, tears sprang to her eyes. Her son tried to comfort her and say things would be all right. She had faith in God, she said.

A few months later, I asked her whether she remembered that conversation. She said she sure did. She didn't sleep that night at home. The image of wearing a bag in order to eat hovered in her mind. "I was horrified," she said.

She recognized that I was trying to be gentle. "But that doesn't change the reality that you knew that another blockage was in the offing." She'd always understood that the ovarian cancer was a looming danger for her, but she really hadn't pictured *how* until then.

She was glad we'd spoken, nonetheless, and so was I. Because the day after her discharge from the hospital, she started vomiting again. The blockage was back. She was readmitted. We put the tube back in.

With a night of fluids and rest, the symptoms once again subsided without need for surgery. But this second episode jolted her because we'd spoken about the meaning of a blockage, that it was her tumor closing in. She saw the connections between events of the previous couple of months, and we talked about the mounting series of crises she'd experienced: the third round of chemotherapy after the previous one had failed, the bad side effects, the pulmonary embolism with its terrible shortness of breath, the bowel obstruction after that, and its almost immediate return. She was starting to grasp that this is what the closing phase of a modern life often looks like—a mounting series of crises from which medicine can offer only brief and temporary rescue. She was experiencing what I have come to think of as the ODTAA syndrome: the syndrome of One Damn Thing After Another. It does not have a totally predictable path. The pauses between crises can vary. But after a certain point, the direction of travel becomes clear.

Douglass did make that trip to Florida. She put her feet in the sand and walked with her husband and saw friends and ate the no-raw-fruits-or-vegetables diet I'd advised her to eat to minimize the chance a fibrous leaf of lettuce got jammed trying to make it through her intestine. Toward the end of the time, she had a fright. She developed bloating after a meal and returned home to Massachusetts a couple days early, worried that the bowel obstruction was back. But the symptoms subsided, and she made a decision. She was going to take a break from her chemotherapy, at least for now. She didn't want to plan her life around the infusions of chemotherapy and the nausea and the painful rashes and the hours of the day she'd spend in bed with fatigue. She wanted to be a wife/mother/neighbor/friend again. She decided, like my father, to take what time would give her, however long that might be.

ONLY NOW DID I begin to recognize how understanding the finitude of one's time could be a gift. After my father was given his diagnosis, he'd initially continued daily life as he always had—his clinical work, his charity projects, his thrice-weekly tennis games—but the sudden knowledge of the fragility of his life narrowed his focus and altered his desires, just as Laura Carstensen's research on perspective suggested it would. It made him visit with his grandchildren more often, put in an extra trip to see his family in India, and tamp down new ventures. He talked about his will with my sister and me and about his plans for sustaining beyond him the college he'd built near his village. One's sense of time can change, though. As the months passed without his symptoms worsening, my father's fear of the future softened. His horizon of time began to lift—it might be years before anything concerning happened, we all thought—and as it did, his ambitions returned. He launched a new construction project for the college in India. He ran for district governor of Rotary for southern Ohio, a position that wouldn't even start for another year, and won the office.

Then, in early 2009, two and a half years after his diagnosis, his symptoms began to change. He developed trouble with his right hand. It started with the tingling and numbness in the tips of his fingers. His grip strength gave out. On the tennis court, the racket began flying out of his hand. He dropped drinking glasses. At work, tying knots and handling catheters grew difficult. With both limbs now developing signs of paralysis, it seemed like he'd come to his line in the sand.

We talked. Wasn't it time for him to stop practicing surgery? And wasn't it time to see Dr. Benzel about surgery for himself?

No, he said. He wasn't ready for either. A few weeks later, however, he announced that he would retire from surgery. As for the spinal operation, he still feared he'd lose more than he'd gain.

After his retirement party that June, I braced for the worst. Surgery had been his calling. It had defined his purpose and meaning in life—his loyalties. He'd wanted to be a doctor since the age of ten, when he saw his young mother die from malaria. So now what was this man going to do with himself?

We witnessed an altogether unexpected transformation. He threw himself into his work as Rotary district governor, whose term of office had just started. He absorbed himself so totally that he changed his e-mail signature from "Atmaram Gawande, M.D." to "Atmaram Gawande, D.G." Somehow, instead of holding on to the lifelong identity that was slipping away from him, he managed to redefine it. He moved his line in the sand. This is what it means to have autonomy—you may not control life's circumstances, but getting to be the author of your life means getting to control what you do with them.

The job of district governor meant spending the year developing the community service work of all the Rotary Clubs in the region. So my father set a goal of speaking at the meetings of each of his district's fifty-nine clubs—twice—and took to the road with my mother. Over the next several months, they crisscrossed a district ten thousand square miles in size. He always did the driving—he could still do that without trouble. They liked to stop at Wendy's for the chicken sandwiches. And he tried to meet as many of the district's thirty-seven hundred Rotarians as he could.

By the following spring, he was completing his second circuit through the district. But the weakness in his left arm had progressed. He couldn't lift it above sixty degrees. His right hand was losing strength, too. And he was starting to have trouble walking. Up until this point, he'd managed to persist with playing tennis but now, to his great dismay, he had to give it up.

"There's a heaviness in my legs," he said. "I'm afraid, Atul."

He and my mother came to visit in Boston. On a Saturday night, the three of us sat in the living room, my mother next to him on a couch and me across from them. I distinctly remember the feeling that a crisis was creeping up on us. He was becoming quadriplegic.

"Is it time for surgery?" I asked him.

"I don't know," he said. It was time, I realized, for our own hard conversation.

"I'm worried," I said. I recalled the list of questions Susan Block, the palliative medicine expert, had said mattered most and posed them to my father one by one. I asked him what his understanding was of what was happening to him.

He understood what I understood. He was becoming paralyzed, he said.

What were his fears if that should happen, I asked?

He said he feared that he would become a burden on my mother and that he wouldn't be able to take care of himself anymore. He couldn't fathom what his life would become. My mother, tearing, said she would be there for him. She would be happy to take care of him. Already the shift had started. He was having her do more and more of the driving, and she arranged his medical appointments now.

What were his goals if his condition worsened, I asked?

He thought on this for a moment. He wanted to finish his Rotary responsibilities, he decided—he would be finishing his term in mid-June. And he wanted to make sure his college and family in India were going to be all right. He wanted to visit them if he could.

I asked him what trade-offs he was willing to make and not willing to make to try to stop what was happening to him. He wasn't sure what I meant. I told him about Susan Block's father, who'd also had a spinal cord tumor. He'd said that if he could still watch football on television and eat chocolate ice cream, that would be good enough for him.

My dad didn't think that would be good enough for him at all. Being with people and interacting with them was what he cared about most, he said. I tried to understand—so even paralysis would be tolerable as long as he could enjoy people's company?

"No," he said. He couldn't accept a life of complete physical paralysis, of needing total care. He wanted to be capable of not only being with people but also still being in charge of his world and life.

His advancing quadriplegia threatened to take that away soon. It would mean twenty-four-hour nursing care, then a ventilator and a feeding tube. He didn't sound like he wanted that, I said.

"Never," he said. "Let me die instead."

Those questions were among the hardest I'd asked in my life. I posed them with great trepidation, fearing, well, I don't know what—anger from my father or mother, or depression, or the sense that just by raising such questions I was letting them down. But what we felt afterward was relief. We felt clarity.

Maybe his answers meant that it *was* time to talk to Benzel about surgery, again, I said. My father softly agreed.

He told Benzel that he was ready for the spinal surgery. He was more afraid now of what the tumor was doing to him than what an operation might do to him. He scheduled the surgery for two months later, after his term of office as district governor ended. By then, his walking had become unsteady. He was having falls and trouble getting up from sitting.

Finally, on June 30, 2010, we arrived at the Cleveland Clinic. My mother, my sister, and I gave him a kiss in a preoperative holding room, adjusted his surgical cap, told him how much we loved him, and left him in the hands of Benzel and his team. The operation was supposed to last all day.

Just two hours into it, however, Benzel came out to the waiting area. He said my father had gone into an abnormal cardiac rhythm. His heart rate sped up to 150 beats a minute. His blood pressure dropped severely. The cardiac monitor showed signs of a potential heart attack, and they halted the operation. With medications, they got him back into a normal rhythm. A cardiologist said his heart rate slowed enough to avoid a full-blown heart attack, but he wasn't sure what had caused the abnormal rhythm. They expected the medications they'd started to prevent its coming back, but there was uncertainty. The operation was not beyond the point of no return. So Benzel had come out to ask us if he should stop or proceed.

I realized then that my father had already told us what to do, just as Susan Block's father had. My dad was more afraid of becoming quadriplegic than of dying. I therefore asked Benzel which posed the greater risk of his becoming quadriplegic in the next couple months: stopping or proceeding? Stopping, he said. We told him to proceed.

He returned seven long hours later. He said my father's heart had remained stable. After the early trouble, all had gone as well as could be hoped. Benzel had been able to perform the decompression procedure successfully and remove a small amount of the tumor, though not more. The back of my father's spine was now open from the top to the bottom of his neck, giving the tumor more room to expand. We'd have to see how he woke up, however, to know if any significant damage had been done.

We sat with my father in the ICU. He was unconscious, on a ventilator. An ultrasound of his heart showed no damage—a huge relief. The team therefore lightened up on his sedatives and let him slowly come to. He woke up groggy but able to follow commands. The resident asked him to squeeze the resident's hands as tightly as he could, to push against him with his feet, to lift his legs off the bed. There was no major loss of motor function, the resident said. When my father heard this, he began gesturing clumsily for our attention. With the breathing tube in his mouth, we couldn't make out what he was saying. He tried to spell what he wanted to say in the air with his finger. L-I-S . . . ? T-A-P . . . ? Was he in pain? Was he having trouble? My sister went through the alphabet and asked him to lift his finger when she got to the right letter. In this way, she deciphered his message. His message was "HAPPY."

A day later he was out of the ICU. Two days after that, he left the hospital for three weeks in a Cleveland rehabilitation facility. He returned home on a hot summer day, feeling strong as ever. He could walk. He had little neck pain at all. He thought trading his old pain for a stiff, unbending neck and a month enduring the hardships of recovery had been a more than acceptable deal. By every measure he'd made the right choices at each step along the way—to put off immediate surgery, to wait even after he'd had to leave his surgical career, to go ahead with the risks only after almost four years, when trouble walking threatened to take away the capabilities he was living for. Soon, he felt, he'd even be able to drive again.

He'd made all the right choices.

THE CHOICES DON'T stop, however. Life is choices, and they are relentless. No sooner have you made one choice than another is upon you.

The results of the tumor biopsy showed my father had an astrocytoma, a relatively slow-growing cancer. After he'd recovered, Benzel referred him to see a radiation oncologist and a neuro-oncologist about the findings. They recommended that he undergo radiation and chemotherapy. This type of tumor cannot be cured, but it can be treated, they said. Treatment could preserve his abilities, perhaps for years, and might even restore some of them. My father was hesitant. He had just recovered and gotten back to his service projects. He was making plans to travel again. He was clear about his priorities, and he was concerned about sacrificing them for yet more treatment. But the specialists pushed him. He had so much to gain from the therapy, they argued, and newer radiation techniques would make the side effects fairly minimal. I pushed him, too. It seemed almost all upside, I said. The primary downside seemed only to be that we had no radiation facility near home capable of providing the treatment. He and my mother would have to move to Cleveland and put their lives on hold for the six weeks of daily radiation treatments. But that was all, I said. He could manage that.

Pressed, he accepted. But how foolish these predictions would turn out to be. Unlike Benzel, the specialists had not been ready to acknowledge how much more uncertain the likelihood of benefit was. Nor had they been ready to take the time to understand my father and what the experience of radiation would be like for him.

At first it seemed like nothing. They'd made a mold of his body for him to lie in so he'd be in the exact same position for each dose of his treatment. He'd lie in the mold for up to an hour, a fishnet mask pulled tight over his face, unable to move two millimeters as the radiation machine clicked and whirred and delivered its daily blast of gamma rays into his brain stem and spinal cord. Over time, however, he experienced stabbing spasms in his back and neck. Each day, the position became harder to endure. The radiation also gradually produced a low-level nausea and a caustic throat pain when he swallowed. With medications, the symptoms became tolerable, but the drugs made him fatigued and constipated. He began sleeping away the day after his treatments, something he'd never done in his life. Then a few weeks into treatment, his sense of taste disappeared. They hadn't mentioned the possibility, and he felt the loss keenly. He loved food. Now he had to force himself to eat.

By the time he returned home, he'd lost twenty-one pounds total. He had a constant tinnitus, a ringing in his ears. His left arm and hand had a new burning, electrical pain. And as for his sense of taste, the doctors expected it to return soon, but it never did.

Nothing improved, in the end. He lost yet more weight that winter. He fell to just 132 pounds. The left-hand numbness and pain climbed above his elbow instead of reducing as hoped. The numbness in his lower extremities rose above his knees. The ringing in his ears was joined by a sense of vertigo. The left side of his face began to droop. The neck and back spasms persisted. He had a fall. A physical therapist recommended a walker, but he didn't want to use it. It felt like failure. The doctors put him on methylphenidate—Ritalin—to try to stimulate his appetite and ketamine, an anesthetic, to control his pain, but the drugs made him hallucinate.

We didn't understand what was happening. The specialists kept expecting the tumor to shrink and, with it, his symptoms. After his six-month MRI, however, he and my mom called me.

"The tumor is expanding," he said, his voice quiet and resigned. The radiation hadn't worked. The images showed that, instead of shrinking, the tumor had kept right on growing, extending upward into his brain, which is why the ringing had persisted and the dizziness had appeared.

I welled with sadness. My mother was angry.

"What was the radiation for?" she asked. "This should have shrunk. They said it would most likely shrink."

My father decided to change the subject. Suddenly, for the first time in weeks, he did not want to talk about his symptoms of the day or his problems. He wanted to know about his grandchildren—how Hattie's symphonic band concert had gone that day, how Walker was doing on his ski team, whether Hunter could say hello. His horizons had narrowed once more.

The doctor recommended seeing the oncologist to plan chemotherapy, and a few days later I joined my parents in Cleveland for the appointment. The oncologist was now center stage, but she too lacked Benzel's ability to take in the whole picture. We missed it keenly. She proceeded in information mode. She laid out eight or nine chemotherapy options in about ten minutes. Average number of syllables per drug: 4.1. It was dizzying. He could take befacizimab, carboplatin, temozolomide, thalidomide, vincristine, vinblastine, or some other options I missed in my notes. She described a variety of different combinations of the drugs to consider as well. The only thing she did not offer or discuss was doing nothing. She suggested he take a combination of temozolomide and befacizimab. She thought that his likelihood of tumor response—that is, of the tumor's not growing further—was around 30 percent. She seemed to not want to sound discouraging, though, so she added that for many patients the tumor becomes "like a low-grade chronic illness" that could be watched.

"You could be back on a tennis court this summer, hopefully," she added.

I couldn't believe she'd really said that. The notion that he might ever get back on a tennis court was daffy—it was not a remotely realistic hope—and I was spitting mad that she would dangle that in front of my father. I saw his expression as he imagined himself back on a tennis court. But it proved to be one of those moments that his being a physician was a clear benefit. He quickly realized it was just a fantasy and, however reluctantly, he turned away from it. Instead, he asked about what the treatment would do to his life.

"Right now, I am foggy in my head. I have tinnitus. I have radiating arm pains. I have trouble walking. That's what's getting me down. Will the drugs make any of this worse?"

She allowed that they could, but it depended on the drug. The discussion became difficult for me or my parents to follow, despite all three of us being doctors. There were too many options, too many risks and benefits to consider with every possible path, and the conversation never got to what he cared about, which was finding a path with the best chance of maintaining a life he'd find worthwhile. She was driving exactly the kind of conversation that I myself tended to have with patients but that I didn't want to have anymore. She was offering data and asking my father to make a choice. Did he want the red pill or the blue pill? But the meaning behind the options wasn't clear at all.

I turned to my mother and father, and said, "Can I ask her about what happens if the tumor progresses?" They nodded. So I did.

The oncologist spoke straightforwardly. His upper extremity weakness would gradually increase, she said. His lower extremity weakness would also advance but respiratory insufficiency—difficulty getting enough oxygen—from the weakness of his chest muscles would become the bigger problem.

Will that feel uncomfortable, my father asked?

No, she said. He'd just grow fatigued and sleepy. But the neck pain and shooting pains would likely increase. He could also develop trouble swallowing as the tumor grew to involve critical nerves.

I asked her what the range of time looked like for people to reach this final point, both with treatment and without.

The question made her squirm. "It's hard to say," she said.

I pushed her. "What's the shortest time you've seen and the longest time you've seen for people who took no treatment?"

Three months was the shortest, she said, three years the longest.

And with treatment?

She got mumbly. Finally she said that the longest might not have been that much more than three years. But with treatment, the average should shift toward the longer end.

It was a hard and unexpected answer for us. "I didn't realize," my father said, his voice trailing off. I remembered what Paul Marcoux, Sara Monopoli's oncologist, had told me about his patients. "I'm thinking, can I get a pretty good year or two out of this? . . . They're thinking ten or twenty years." We were thinking ten or twenty years, too.

My father decided to take some time to consider his options. She gave him a prescription for a steroid pill that might temporarily slow the tumor's growth, while having relatively few side effects. That night, my parents and I went out for dinner.

"The way things are going I could be bedridden in a few months," my father said. The radiation therapy had only made matters worse. Suppose chemotherapy did the same? We needed guidance. He was torn between living the best he could with what he had versus sacrificing the life he had left for a murky chance of time later.

One of the beauties of the old system was that it made these decisions simple. You took the most aggressive treatment available. It wasn't a decision at all, really, but a default setting. This business of deliberating on your options—of figuring out your priorities and working with a doctor to match your treatment to them—was exhausting and complicated, particularly when you didn't have an expert ready to help you parse the unknowns and ambiguities. The pressure remains all in one direction, toward doing more, because the only mistake clinicians seem to fear is doing too little. Most have no appreciation that equally terrible mistakes are possible in the other direction—that doing too much could be no less devastating to a person's life.

My father went home still uncertain what to do. Then he had a series of five or six falls. The numbness in his legs was getting worse. He began losing the sense of where his feet were underneath him. One time, going down, he hit his head hard and had my mother call 911. The EMTs arrived, siren wailing. They put him on a backboard and in a hard collar and raced him to the ER. Even in his own hospital, it was three hours before he could get the X-rays confirming that nothing was broken and that he could sit up and take the collar off. By then, the stiff collar and rock-hard backboard had put him in excruciating pain. He required multiple injections of morphine to control it and wasn't released home until near midnight. He told my mother he never wanted to be put through that kind of experience again.

Two mornings later, I got a call from my mother. Around 2:00 a.m., my father had gotten out of bed to go to the bathroom, she said, but when he went to stand up, his legs wouldn't hold him, and he went down. The floor was carpeted. He didn't hit his head and didn't seem hurt. But he couldn't get himself up. His arms and legs were too weak. She tried to lift him back into bed, but he was too heavy. He didn't want to call an ambulance again. So they decided to wait until morning for help. She pulled blankets and pillows off the bed for him and lay down beside him, not wanting him to be alone. But with her bad arthritic knees—she was seventy-five years old herself—she found she now couldn't get up either. Around 8:00 a.m., the housekeeper arrived and found them both on the floor. She helped my mother to her feet and my father into bed. That was when my mother called. She sounded frightened. I asked her to put my dad on the line. He was crying, frantic, sputtering, hard to understand.

"I'm so scared," he said. "I'm becoming paralyzed. I can't do this. I don't want this. I don't want to go through this. I want to die rather than go through this."

Tears wet my eyes. I'm a surgeon. I like solving things. But how do I solve this? For two minutes, I tried to just listen as he repeated over and over that he couldn't do this. He asked me if I could come.

"Yes," I said.

"Can you bring the kids?" He thought he was dying. But the hard thing was that he was not. He could be this way for a long while, I realized.

"Let me come first," I told him.

I set about arranging a plane ticket back home to Ohio and

canceling my patients and commitments in Boston. Two hours later he called back. He'd calmed down. He'd been able to stand up again, even walk to the kitchen. "You don't have to come," he said. "Come on the weekend." But I decided to go; the crises were mounting.

When I made it to Athens early that evening, my mother and father were sitting at the dinner table eating, and they had already turned the six hours he spent paralyzed on the bedroom floor into a comedy in the retelling.

"It's been years since I've been down on the floor," my mother said.

"It was almost romantic," my father said, with what I can only describe as a giggle.

I tried to roll with it. But the person I saw before me was different from the one I'd seen just a few weeks before. He'd lost more weight. He was so weak his speech sometimes slurred. He had trouble getting food into his mouth, and his shirt was smeared with his dinner. He needed help standing from sitting. He'd become old before my eyes.

Trouble was coming. Today was the first day I really grasped what it would mean for him to become paralyzed. It meant difficulty with the basics—standing up, getting to the bathroom, getting bathed, getting dressed—and my mother wasn't going to be able to help him. We needed to talk.

Later that night, I sat with my parents and asked, "What are we going to do to take care of you, Dad?"

"I don't know," he said.

"Have you had trouble getting your breath?"

"He can breathe," my mom said.

"We're going to need a proper way to take care of him," I said to her.

"Maybe they can give him chemo," she said.

"No," he said sharply. He'd made up his mind. Even just the side effects of the steroids were proving difficult for him to tolerate—sweats, anxiety, difficulties with thinking and moodiness—and he'd recognized no benefit. He did not think a full-blown course of chemotherapy was going to make any radical improvement, and he did not want the side effects.

I helped my mother get him to bed when it got late. I talked with her about the help he was going to need. He was going to need nursing care, a hospital bed, an air mattress to prevent bedsores, physical therapy to prevent his muscles from stiffening. Should we look at nursing homes?

She was aghast. Absolutely not, she said. She'd had friends in the ones around town, and they'd appalled her. She could not imagine putting him in any of them.

We'd come to the same fork in the road I have seen scores of patients come to, the same place I'd seen Alice Hobson come to. We were up against the unfixable. But we were desperate to believe that we weren't up against the unmanageable. Yet short of calling 911 the next time trouble hit, and letting the logic and momentum of medical solutions take over, what were we to do? Between the three of us we had 120 years of experience in medicine, but it seemed a mystery. It turned out to be an education.

WE NEEDED OPTIONS, and Athens was not a place where anyone could expect the kinds of options for the frail and aged that I'd seen sprouting in Boston. It is a small town in the foothills of Appalachia. The local college, Ohio University, is its lifeblood. One-third of the county lived in poverty, making ours the poorest county in the state. So it seemed a surprise when I asked around and discovered that even here people were rebelling against the way medicine and institutions take control of their lives in old age.

I spoke, for instance, to Margaret Cohn. She and her husband, Norman, were retired biologists. He had a severe form of arthritis known as ankylosing spondylitis and, because of a tremor and the effects of a polio infection in his youth, he faced increasing difficulty walking. The two of them were becoming concerned about whether they'd be able to manage in their home on their own. They didn't want to be forced to move in with any of their three children, who were scattered far away. They wanted to stay in the community. But when they looked around town for assisted living options, nothing was remotely acceptable. "I would live in a tent before I would live like that," she told me.

She and Norman decided to come up with a solution themselves, their age be damned. "We realized, if we didn't do it, no one was going to do it for us," she said. Margaret had read an article in the newspaper about Beacon Hill Village, the Boston program that created neighborhood support for the aged to stay in their homes, and she was inspired. The Cohns got a group of friends together, and in 2009 they formed Athens Village on the same model. They calculated that, if they could get seventy-five people to pay four hundred dollars per year, it would be enough to establish the essential services. A hundred people signed up, and Athens Village was under way.

One of the first people they hired was a wonderfully friendly handyman. He was willing to help people with all the mundane household matters that you take for granted when you're able but that become critical to surviving in your home when you're not—fixing a broken lock, changing a lightbulb, sorting out what to do about a broken water heater.

"He could do almost anything. People who joined felt the maintenance guy alone was worth the four hundred dollars," Margaret said.

They also hired a part-time director. She checked up on people and pulled together volunteers who could stop by if the power was out or someone needed a casserole. A local visiting nurse agency provided free office space and a member discount on nursing aide costs. Church and civic organizations provided a daily van transportation service and meals-on-wheels for members who needed it. Bit by bit, Athens Village built services and a community that could ensure that members were not left flailing when their difficulties mounted. It came not a moment too soon for the Cohns. A year after they'd founded it, Margaret took a fall that put her permanently in a wheelchair. Even with both of them disabled and in their mideighties, they were able to make staying at home work.

My parents and I talked about joining Athens Village. The only other option was home hospice care, and I hesitated to raise it. Its mere mention would drag the dark, unspoken subject of dying onto the coffee table between us. Discussing Athens Village let us pretend what my father was going through was just a kind of aging. But I steeled myself and asked whether home hospice was something to consider, as well.

My father, it turned out, was willing to contemplate hospice, my mother less so. "I don't think it's necessary," she said. But my father said that maybe it wasn't a bad idea to have someone from the agency tell us about it.

The next morning a nurse practitioner from Appalachian Community Hospice stopped by. My mother made some tea, and we sat around our dining table. I will confess to expecting little of the nurse. This wasn't Boston. The agency was called Appalachian Community Hospice, for God's sake. The nurse blew me away, though.

"How are you?" she said to my dad. "Do you have a lot of pain?"

"Not right now," he said.

"Where do you get the pain?"

"In my neck and in my back."

With that opening, I realized, she had established a few things. She'd made sure he was in a state of mind to talk. She'd made instantly clear that what she cared about was him and how he was doing, not about his disease or his diagnosis. And she'd let us know that, surrounded by a bunch of doctors or not, she knew exactly what she was doing.

She looked to be around fifty, with short, cropped gray hair, a white cotton sweater with an embroidered rose across the front, and a stethoscope sticking out of her bag. She had a local, country accent. And with it, she got right to the point.

"They sent me out with hospice papers," she said to my father. "What do you think about that?"

My father didn't say anything for a moment. The nurse waited. She knew how to be silent.

"I think it may be best," he said, "because I don't want chemo."

"What kinds of problems are you having?"

"Nausea," he said. "Pain control. Grogginess. The medicine makes me too sleepy. I've tried Tylenol with codeine. I've tried Toradol pills. Now I'm on ketamine."

He went on. "I woke up this morning and it was a big change. I couldn't stand up. I couldn't push the pillow up in the bed. I couldn't handle a toothbrush to brush my teeth. I couldn't pull my pants or socks on. My torso is becoming weak. It's getting hard to sit up."

"Hospice is about palliative care," she said, about giving care to help manage these difficulties. She went through the services that Medicare would cover for my father. He'd have a palliative care physician who could help adjust medications and other treatments to minimize his nausea, pain, and other symptoms as much as possible. He'd have regular nursing visits plus emergency nursing support available twenty-four hours a day by phone. He'd have fourteen hours a week of a home health aide, who could help with bathing, getting dressed, cleaning up the house, anything nonmedical. There'd be a social worker and spiritual counselor available. He'd have the medical equipment he needed. And he could "revocate"—drop the hospice services—at any time.

She asked him if these were services he'd like to start now or think about.

"Start now," he said. He was ready. I looked at my mother. Her face was blank.

The nurse practitioner got into the nitty-gritty: Did he have a DNR? A baby monitor or a bell for him to summon a caregiver? A 24-7 presence in the house to help?

Then she asked, "What funeral home do you want to use?" and I was divided between shock—are we really having this conversation?—and reassurance at how normal and routine this was to her.

"Jagers," he said, without hesitation. He'd been thinking about it all along, I realized. My father was calm. My mother, however, was stunned. This was not going where she'd been prepared for it to go.

The nurse turned to her and, not unkindly but nonetheless all too clearly, said, "When he passes away, don't call 911. Don't call the police. Don't call an ambulance company. Call us. A nurse will help. She will discard the narcotics, arrange the death certificate, wash his body, arrange with the funeral home."

"Right now, we're not thinking of death," my mother said firmly. "Just paralysis."

"Okay," the nurse said.

She asked my father what his biggest concerns were. He said

he wanted to stay strong while he could. He wanted to be able to type, because e-mail and Skype were how he connected with family and friends all over the world. He didn't want pain.

"I want to be happy," he said.

She stayed almost two hours. She examined him, inspected the house for hazards, sorted out where to place the bed, and figured out a schedule for the nurse and the home health aide to visit. She also told my father he needed to do just two main things. She figured out he'd been taking his pain medications haphazardly, tinkering with which drug he took at what dose, and she told him he needed to take a consistent regimen and log his response so the hospice team could gauge the effect accurately and help him find the optimal mix to minimize pain and grogginess. And she told him that he needed to no longer attempt to get up or around without someone helping him.

"I'm used to just getting up and walking," he said.

"If you break your hip, Dr. Gawande, it will be a disaster," she said.

He agreed to her instructions.

In the days that followed, it astonished me to see the difference the hospice's two simple instructions made. My father couldn't resist still tinkering with his medications, but he did it much less than he had and he kept a log of his symptoms and what meds he took when. The nurse who visited each day would go over it with him and identify adjustments to make. He'd been oscillating wildly, we realized, between severe pain and becoming so drugged he seemed drunk, with slurred, confused speech and difficulty controlling his limbs. The changes gradually smoothed the pattern. The drunken episodes all but disappeared. And his pain control improved, although it was never complete, to his great frustration and sometimes anger.

He also complied with the instructions not to attempt to get around without help. The hospice helped my parents hire a personal care aide to stay overnight and assist my father to the bathroom when he needed it. After that, he had no more falls, and we gradually realized how much each one had set him back. Every passing day without a fall allowed his back and neck spasms to reduce, his pain to become better controlled, and his strength to increase.

We witnessed for ourselves the consequences of living for the best possible day today instead of sacrificing time now for time later. He'd become all but wheelchair bound. But his slide into complete quadriplegia halted. He became more able to manage short distances with a walker. His control of his hands and his arm strength improved. He had less trouble calling people on the phone and using his laptop. The greater predictability of his day let him have more visitors over. Soon he even began hosting parties at our house again. He found that in the narrow space of possibility that his awful tumor had left for him there was still room to live.

Two months on, in June, I flew home from Boston not only to see him but also to give the graduation address for Ohio University. My father had been excited about attending the convocation from the moment I had been invited a year before. He was proud, and I had envisioned both my parents being there. Little is more gratifying than actually being wanted back in your hometown. For a while, however, I feared my father might not survive long enough. In the last few weeks, it became apparent he would, and the planning turned to logistics.

The ceremony was to take place in the university's basketball arena with the graduates in folding chairs on the parquet and their families up in the stands. We worked out a plan to bring my father up the outside ramp by golf cart, transfer him to a wheelchair, and seat him on the periphery of the floor to watch.

But when the day came and the cart brought him to the arena doors, he was adamant that he would walk and not sit in a wheelchair on the floor.

I helped him to stand. He took my arm. And he began walking. I'd not seen him make it farther than across a living room in half a year. But walking slowly, his feet shuffling, he went the length of a basketball floor and then up a flight of twenty concrete steps to join the families in the stands. I was almost overcome just witnessing it. Here is what a different kind of care—a different kind of medicine—makes possible, I thought to myself. Here is what having a hard conversation can do.

8 · Courage

In 380 BC, Plato wrote a dialogue, the *Laches*, in which Socrates and two Athenian generals seek to answer a seemingly simple question: What is courage? The generals, Laches and Nicias, had gone to Socrates to resolve a dispute between them over whether boys undergoing military training should be taught to fight in armor. Nicias thinks they should. Laches thinks they shouldn't.

Well, what's the ultimate purpose of the training? Socrates asks.

To instill courage, they decide.

So then, "What is courage?"

Courage, Laches responds, "is a certain endurance of the soul."

Socrates is skeptical. He points out that there are times when the courageous thing to do is not to persevere but to retreat or even flee. Can there not be foolish endurance?

Laches agrees but tries again. Perhaps courage is "wise endurance."

This definition seems more apt. But Socrates questions whether courage is necessarily so tightly joined to wisdom. Don't we admire courage in the pursuit of an unwise cause, he asks?

Well, yes, Laches admits.

Now Nicias steps in. Courage, he argues, is simply "knowledge of what is to be feared or hoped, either in war or in anything else." But Socrates finds fault here, too. For one can have courage without perfect knowledge of the future. Indeed, one often must.

The generals are stumped. The story ends with them coming to no final definition. But the reader comes to a possible one: Courage is *strength* in the face of knowledge of what is to be feared or hoped. Wisdom is prudent strength.

At least two kinds of courage are required in aging and sickness. The first is the courage to confront the reality of mortality—the courage to seek out the truth of what is to be feared and what is to be hoped. Such courage is difficult enough. We have many reasons to shrink from it. But even more daunting is the second kind of courage—the courage to act on the truth we find. The problem is that the wise course is so frequently unclear. For a long while, I thought that this was simply because of uncertainty. When it is hard to know what will happen, it is hard to know what to do. But the challenge, I've come to see, is more fundamental than that. One has to decide whether one's fears or one's hopes are what should matter most.

I HAD RETURNED to Boston from Ohio, and to my work at the hospital, when I got a late-night page: Jewel Douglass was back, unable to hold food down again. Her cancer was progressing. She'd made it three and a half months—longer than I'd thought she'd have, but shorter than she'd expected. For a week, the symptoms had mounted: they started with bloating, became waves of crampy abdominal pain, then nausea, and progressed to vomiting. Her oncologist sent her to the hospital. A scan showed her ovarian cancer had multiplied, grown, and partly obstructed her intestine again. Her abdomen had also filled with fluid, a new problem for her. The deposits of tumor had stuffed up her lymphatic system, which serves as a kind of storm drain for the lubricating fluids that the body's internal linings secrete. When the system is blocked, the fluid has nowhere to go. When that happens above the diaphragm, as it did with Sara Monopoli's lung cancer, the chest fills up like a ribbed bottle until you have trouble breathing. If the system gets blocked up below the diaphragm, as it did with Douglass, the belly fills up like a rubber ball until you feel as if you will burst.

Walking into Douglass's hospital room, I'd never have known she was as sick as she was if I hadn't seen the scan. "Well, look who's here!" she said, as if I'd just arrived at a cocktail party. "How are you, doctor?"

"I think I'm supposed to ask you that," I said.

She smiled brightly and pointed around the room. "This is my husband, Arthur, whom you know, and my son, Brett." She got me grinning. Here it was eleven o'clock at night, she couldn't hold down an ounce of water, and still she had her lipstick on, her silver hair brushed straight, and she was insisting on making introductions. She wasn't oblivious to her predicament. She just hated being a patient and the grimness of it all.

I talked to her about what the scan showed. She had no unwillingness to face the facts. But what to do about them was another matter. Like my father's doctors, the oncologist and I had a menu of options. There was a whole range of new chemotherapy regimens that could be tried to shrink the tumor burden. I had a few surgical options for dealing with her situation, as well. With surgery, I told her, I wouldn't be able to remove the intestinal blockage, but I might be able to bypass it. I'd either connect an obstructed loop to an unobstructed one or I'd disconnect the bowel above the blockage and give her an ileostomy, which she'd have to live with. I'd also put in a couple drainage catheters—permanent spigots that could be opened to release the fluids from her blocked-up drainage ducts or intestines when necessary. Surgery risked serious complications—wound breakdown, leakage of bowel into her abdomen, infections—but it offered her the only way she could regain her ability to eat. I also told her that we did not have to do either chemo or surgery. We could provide medications to control her pain and nausea and arrange for hospice at home.

The options overwhelmed her. They all sounded terrifying. She didn't know what to do. I realized, with shame, that I'd reverted to being Dr. Informative—here are the facts and figures; what do you want to do? So I stepped back and asked the questions I'd asked my father: What were her biggest fears and concerns? What goals were most important to her? What tradeoffs was she willing to make, and what ones was she not?

Not everyone is able to answer such questions, but she did. She said she wanted to be without pain, nausea, or vomiting. She wanted to eat. Most of all, she wanted to get back on her feet. Her biggest fear was that she wouldn't be able to live life again and enjoy it—that she wouldn't be able to return home and be with the people she loved.

As for what trade-offs she was willing to make, what sacrifices she was willing to endure now for the possibility of more time later, "Not a lot," she said. Her perspective on time was shifting, focusing her on the present and those closest to her. She told me that uppermost in her mind was a wedding that weekend that she was desperate not to miss. "Arthur's brother is marrying my best friend," she said. She'd set them up on their first date. Now the wedding was just two days away, on Saturday at 1:00 p.m. "It's just the *best* thing," she said. Her husband was going to be the ring bearer. She was supposed to be a brides-maid. She was willing to do anything to be there, she said.

The direction suddenly became clear. Chemotherapy had only a slim chance of improving her current situation and it came at substantial cost to the time she had now. An operation would never let her get to the wedding, either. So we made a plan to see if we could get her there. We'd have her come back afterward to decide on the next steps.

With a long needle, we tapped a liter of tea-colored fluid from her abdomen, which made her feel at least temporarily better. We gave her medication to control her nausea. And she was able to drink enough liquids to stay hydrated. At three o'clock Friday afternoon, we discharged her with instructions to drink nothing thicker than apple juice and return to see me after the wedding.

She didn't make it. She came back to the hospital that same night. Just the car ride, with all its swaying and bumps, set her vomiting again. The crampy attacks returned. Things only got worse at home.

We agreed surgery was the best course now and scheduled her for it the next day. I would focus on restoring her ability to eat and putting drainage tubes in. Afterward, she could decide if she wanted more chemotherapy or to go on hospice. She was as clear as I've seen anyone be about her goals and what she wanted to do to achieve them.

Yet still she was in doubt. The following morning, she told me to cancel the operation.

"I'm afraid," she said. She didn't think she had the courage to go ahead with the procedure. She'd tossed all night thinking about it. She imagined the pain, the tubes, the indignities of the possible ileostomy, and then there were the incomprehensible horrors of the complications she could face. "I don't want to take risky chances," she said.

As we talked, it became clear that her difficulty wasn't lack of courage to act in the face of risks. Her difficulty was in sorting out how to think about them. Her greatest fear was of suffering, she said. Although we were doing the operation in order to reduce her suffering, couldn't the procedure make it worse rather than better?

Yes, I said. It could. Surgery offered her the possibility of being able to eat again and a very good likelihood of controlling her nausea, but it carried substantial risk of giving her only pain without improvement or adding yet new miseries. She had, I estimated for her, a 75 percent chance I'd make her future better, at least for a little while, and a 25 percent chance I'd make it worse.

So what then was the right thing for her to do? And why was the choice so agonizing? The choice, I realized, was far more complicated than a risk calculation. For how do you weigh relief from nausea, and the chances of being able to eat again, against the possibilities of pain, of infections, of having to live with stooling into a bag?

The brain gives us two ways to evaluate experiences like suffering—there is how we apprehend such experiences in the moment and how we look at them afterward—and the two ways are deeply contradictory. The Nobel Prize–winning researcher Daniel Kahneman illuminated what happens in a series of experiments he recounted in his seminal book *Thinking, Fast and Slow*. In one of them, he and University of Toronto physician Donald Redelmeier studied 287 patients undergoing colonoscopy and kidney stone procedures while awake. The researchers gave the patients a device that let them rate their pain every sixty seconds on a scale of one (no pain) to ten (intolerable pain), a system that provided a quantifiable measure of their moment-by-moment experience of suffering. At the end, the patients were also asked to rate the total amount of pain they experienced during the procedure. The procedures lasted anywhere from four minutes to more than an hour. And the patients typically reported extended periods of low to moderate pain punctuated by moments of significant pain. A third of the colonoscopy patients and a quarter of the kidney stone patients reported a pain score of ten at least once during the procedure.

Our natural assumption is that the final ratings would represent something like the sum of the moment-by-moment ones. We believe that having a longer duration of pain is worse than a shorter duration and that having a greater average level of pain is worse than having a lower average level. But this wasn't what the patients reported at all. Their final ratings largely ignored the duration of pain. Instead, the ratings were best predicted by what Kahneman termed the "Peak-End rule": an average of the pain experienced at just two moments—the single worst moment of the procedure and the very end. The gastroenterologists conducting the procedures rated the level of pain they had inflicted very similarly to their patients, according to the level of pain at the moment of greatest intensity and the level at the end, not according to the total amount.

People seemed to have two different selves—an experiencing self who endures every moment equally and a remembering self who gives almost all the weight of judgment afterward to two single points in time, the worst moment and the last one. The remembering self seems to stick to the Peak-End rule even when the ending is an anomaly. Just a few minutes without pain at the end of their medical procedure dramatically reduced patients' overall pain ratings even after they'd experienced more than half an hour of high level of pain. "That wasn't so terrible," they'd reported afterward. A bad ending skewed the pain scores upward just as dramatically.

Studies in numerous settings have confirmed the Peak-End rule and our neglect of duration of suffering. Research has also shown that the phenomenon applies just as readily to the way people rate pleasurable experiences. Everyone knows the experience of watching sports when a team, having performed beautifully for nearly the entire game, blows it in the end. We feel that the ending ruins the whole experience. Yet there's a contradiction at the root of that judgment. The experiencing self had whole hours of pleasure and just a moment of displeasure, but the remembering self sees no pleasure at all.

If the remembering self and the experiencing self can come to radically different opinions about the same experience, then the difficult question is which one to listen to. This was Jewel Douglass's torment at bottom, and to a certain extent mine, if I was to help guide her. Should we listen to the remembering—or, in this case, anticipating—self that focuses on the worst things she might endure? Or should we listen to the experiencing self, which would likely have a lower average amount of suffering in the time to come if she underwent surgery rather than if she just went home—and might even get to eat for a while again?

In the end, people don't view their life as merely the average of all of its moments—which, after all, is mostly nothing much plus some sleep. For human beings, life is meaningful because it is a story. A story has a sense of a whole, and its arc is determined by the significant moments, the ones where something happens. Measurements of people's minute-by-minute levels of pleasure and pain miss this fundamental aspect of human existence. A seemingly happy life may be empty. A seemingly difficult life may be devoted to a great cause. We have purposes larger than ourselves. Unlike your experiencing self—which is absorbed in the moment—your remembering self is attempting to recognize not only the peaks of joy and valleys of misery but also how the story works out as a whole. That is profoundly affected by how things ultimately turn out. Why would a football fan let a few flubbed minutes at the end of the game ruin three hours of bliss? Because a football game is a story. And in stories, endings matter.

Yet we also recognize that the experiencing self should not be ignored. The peak and the ending are not the only things that count. In favoring the moment of intense joy over steady happiness, the remembering self is hardly always wise.

"An inconsistency is built into the design of our minds," Kahneman observes. "We have strong preferences about the duration of our experiences of pain and pleasure. We want pain to be brief and pleasure to last. But our memory . . . has evolved to represent the most intense moment of an episode of pain or pleasure (the peak) and the feelings when the episode was at its end. A memory that neglects duration will not serve our preference for long pleasure and short pains."

When our time is limited and we are uncertain about how best to serve our priorities, we are forced to deal with the fact that both the experiencing self and the remembering self matter. We do not want to endure long pain and short pleasure. Yet certain pleasures can make enduring suffering worthwhile. The peaks are important, and so is the ending.

Jewel Douglass didn't know if she was willing to face the suffering that surgery might inflict on her and feared being left worse off. "I don't want to take risky chances," she said, and by that, I realized, she meant that she didn't want to take a high-stakes gamble on how her story would turn out. On the one hand, there was so much she still hoped for, however seemingly mundane. That very week, she'd gone to church, driven to the store, made family dinner, watched a television show with

Arthur, had her grandson come to her for advice, and made wedding plans with dear friends. If she could be allowed to have even a little of that—if she could be freed from what her tumor was doing to her to enjoy just a few more such experiences with the people she loved—she would be willing to endure a lot. On the other hand, she didn't want to chance a result even worse than the one she already faced with her intestines cinched shut and fluid filling her abdomen like a dripping faucet. It seemed as if there were no way forward. But as we talked that Saturday morning in her hospital room, with her family around her and the operating room standing by downstairs, I came to understand she was telling me everything I needed to know.

We should go to surgery, I told her, but with the directions she'd just spelled out—to do what I could to enable her to return home to her family while not taking risky chances. I'd put in a small laparoscope. I'd look around. And I'd attempt to unblock her intestine only if I saw that I could do it fairly easily. If it looked difficult and risky, then I'd just put in tubes to drain her backed-up pipes. I'd aim to do what might have sounded like a contradiction in terms: a palliative operation, an operation whose overriding priority, whatever the violence and risks inherent, was to do only what was likely to make her feel better immediately.

She remained quiet, thinking.

Her daughter took her hand. "We should do this, Mom," she said.

"Okay," Douglass said. "But no risky chances."

"No risky chances," I said.

When she was asleep under anesthesia, I made a half-inch incision above her belly button. It let out a gush of thin, blood-tinged fluid. I slipped my gloved finger inside to feel for space to insert the fiberoptic scope. But a hard loop of tumor-caked bowel blocked the entry. I wasn't even going to be able to put in a camera. I had the resident take the knife and extend the incision upward until it was large enough to see in directly and get a hand inside. At the bottom of the hole, I saw a free loop of distended bowel—it looked like an overinflated pink inner tube—that I thought we might be able to pull up to the skin and make into an ileostomy so she could eat again. But it remained tethered by tumor, and as we tried to chip it free it became evident that we were risking creating holes we'd never be able to repair. Leakage inside the abdomen would be a calamity. So we stopped. Her aims for us were clear. No risky chances. We shifted focus and put in two long, plastic drainage tubes. One we inserted directly into her stomach in order to empty the contents backed up there; the other we laid in the open abdominal cavity to empty the fluid outside her gut. Then we closed up, and we were done.

I told her family we weren't able to help her eat again, and when Douglass woke up I told her, as well. Her daughter had tears. Her husband thanked us for trying. Douglass tried to put a brave face on it.

"I was never obsessed with food anyway," she said.

The tubes relieved her nausea and abdominal pain greatly—"90 percent," she said. The nurses taught her how to open the gastric tube into a bag when she felt sick and the abdominal tube when her belly felt too tight. We told her she could drink whatever she wanted and even eat soft food for the taste. Three days after surgery she went home with hospice to look after her. Before she left, her oncologist and the oncology nurse practitioner saw her. Douglass asked them how long they thought she had.

"They both filled up with tears," she told me. "It was kind of my answer."

A few days after Douglass left the hospital, she and her family allowed me to stop by her home after work. She answered the door herself, wearing a robe because of the tubes and apologizing for it. We sat in her living room, and I asked how she was doing.

She was doing okay, she said. "I think I have a measure that I'm slip, slip, slipping," but she had been seeing old friends and relatives all day, and she loved it. "It's my lifeblood, really, so I want to do it." Her family staggered the visits to keep them from tiring her out.

She said she didn't like all the contraptions sticking out of her. The tubes were uncomfortable where they poked out of her belly. "I didn't know that there would be this constant pressure," she said. But the first time she found that just opening a tube could take away her nausea, "I looked at the tube and said, 'Thank you for being there.'"

She was taking just Tylenol for pain. She didn't like narcotics because they made her drowsy and weak, and that interfered with seeing people. "I've probably confused the hospice people because I said at some point, 'I don't want any discomfort. Bring it on'"—by which she meant the narcotics. "But I'm not there yet."

Mostly, we talked about memories from her life, and they were good ones. She was at peace with God, she said. I left feeling that, at least this once, we'd learned to do it right. Douglass's story was not ending the way she ever envisioned, but it was nonetheless ending with her being able to make the choices that meant the most to her.

Two weeks later, her daughter, Susan, sent me a note. "Mom died on Friday morning. She drifted quietly to sleep and took her last breath. It was very peaceful. My dad was alone by her side with the rest of us in the living room. This was such a perfect ending and in keeping with the relationship they shared."

I AM LEERY of suggesting the idea that endings are controllable. No one ever really has control. Physics and biology and accident ultimately have their way in our lives. But the point is that we are not helpless either. Courage is the strength to recognize *both* realities. We have room to act, to shape our stories, though as time goes on it is within narrower and narrower confines. A few conclusions become clear when we understand this: that our most cruel failure in how we treat the sick and the aged is the failure to recognize that they have priorities beyond merely being safe and living longer; that the chance to shape one's story is essential to sustaining meaning in life; that we have the opportunity to refashion our institutions, our culture, and our conversations in ways that transform the possibilities for the last chapters of everyone's lives.

Inevitably, the question arises of how far those possibilities should extend at the very end—whether the logic of sustaining people's autonomy and control requires helping them to accelerate their own demise when they wish to. "Assisted suicide" has become the term of art, though advocates prefer the euphemism "death with dignity." We clearly already recognize some form of this right when we allow people to refuse food or water or medications and treatments, even when the momentum of medicine fights against it. We accelerate a person's demise every time we remove someone from an artificial respirator or artificial feeding. After some resistance, cardiologists now accept that patients have the right to have their doctors turn off their pacemaker—the artificial pacing of their heart—if they want it. We also recognize the necessity of allowing doses of narcotics and sedatives that reduce pain and discomfort even if they may knowingly speed death. All proponents seek is the ability for suffering people to obtain a prescription for the same kind of medications, only this time to let them hasten the timing of their death. We are running up against the difficulty of maintaining a coherent philosophical distinction between giving people the right to stop external or artificial processes that prolong their lives and giving them the right to stop the natural, internal processes that do so.

At root, the debate is about what mistakes we fear most—the mistake of prolonging suffering or the mistake of shortening valued life. We stop the healthy from committing suicide because we recognize that their psychic suffering is often temporary. We believe that, with help, the remembering self will later see matters differently than the experiencing self—and indeed only a minority of people saved from suicide make a repeated attempt; the vast majority eventually report being glad to be alive. But for the terminally ill who face suffering that we know will increase, only the stonehearted can be unsympathetic.

All the same, I fear what happens when we expand the terrain of medical practice to include actively assisting people with speeding their death. I am less worried about abuse of these powers than I am about dependence on them. Proponents have crafted the authority to be tightly circumscribed to avoid error and misuse. In places that allow physicians to write lethal prescriptions—countries like the Netherlands, Belgium, and Switzerland and states like Oregon, Washington, and Vermont—they can do so only for terminally ill adults who face unbearable suffering, who make repeated requests on separate occasions, who are certified not to be acting out of depression or other mental illness, and who have a second physician confirming they meet the criteria. Nonetheless, the larger culture invariably determines how such authority is employed. In the Netherlands, for instance, the system has existed for decades, faced no serious opposition, and significantly grown in use. But the fact that, by 2012, one in thirty-five Dutch people sought assisted suicide at their death is not a measure of success. It is a measure of failure. Our ultimate goal, after all, is not a good death but a good life to the very end. The Dutch have been slower than others to develop palliative care programs that might provide for it. One reason, perhaps, is that their system of assisted death may have reinforced beliefs that reducing suffering and improving lives through other means is not feasible when one becomes debilitated or seriously ill.

Certainly, suffering at the end of life is sometimes unavoidable and unbearable, and helping people end their misery may be necessary. Given the opportunity, I would support laws to provide these kinds of prescriptions to people. About half don't even use their prescription. They are reassured just to know they have this control if they need it. But we damage entire societies if we let providing this capability divert us from improving the lives of the ill. Assisted living is far harder than assisted death, but its possibilities are far greater, as well.

In the throes of suffering, this can be difficult to see. One day I got a call from the husband of Peg Bachelder, my daughter Hunter's piano teacher. "Peg's in the hospital," Martin said.

I'd known she had serious health issues. Two and a half years earlier, she'd developed a right hip pain. The condition was misdiagnosed for almost a year as arthritis. When it got worse, one physician even recommended seeing a psychiatrist and gave her a book on "how to let go of your pain." But imaging finally revealed that she had a five-inch sarcoma, a rare soft-tissue cancer, eating into her pelvis and causing a large blood clot in her leg. Treatment involved chemotherapy, radiation, and radical surgery removing a third of her pelvis and reconstructing it with metal. It was a year in hell. She was hospitalized for months with complications. She'd loved cycling, yoga, walking her Shetland sheepdog with her husband, playing music, and teaching her beloved students. She'd had to let go of all of that.

Eventually, however, Peg recovered and was able to return to teaching. She needed Canadian crutches—the kind that have a cuff around the forearm—to get around but otherwise remained her graceful self and refilled her roster of students in no time. She was sixty-two, tall, with big round glasses, a thick bob of auburn hair, and a lovely gentle way that made her an immensely popular teacher. When my daughter struggled with grasping a sound or technique, Peg was never hurried. She'd have her try this and then try that, and when Hunter finally got it, Peg would burble with genuine delight and hug her close.

A year and a half after returning, Peg was found to have a leukemia-like malignancy caused by her radiation treatment. She went back on chemotherapy but somehow kept teaching through it. Every few weeks, she'd have to reschedule Hunter's lesson, and we had to explain the situation to Hunter, who was just thirteen at the time. But Peg always found a way to keep going.

Then for two straight weeks, she postponed the lessons. That was when I got the call from Martin. He was phoning from the hospital. Peg had been admitted for several days. He put his cell on speaker so she could talk. She sounded weak—there were long pauses when she spoke—but she was clear-voiced about the situation. The leukemia treatment had stopped working a few weeks before, she said. She developed a fever and infection due to her compromised immune system. Imaging also showed her original cancer had come back in her hip and in her liver. The recurrent disease began to cause immobilizing hip pain. When it made her incontinent, that felt like the final straw. She checked into the hospital at that point, and she didn't know what to do.

What had the doctors told her they could do? I asked.

"Not much," she said. She sounded flat, utterly hopeless. They were giving her blood transfusions, pain medications, and steroids for tumor-caused fevers. They'd stopped giving her chemotherapy.

I asked her what her understanding of her condition was.

She said she knew she was going to die. There's nothing more they can do, she said, an edge of anger creeping into her voice.

I asked her what her goals were, and she didn't have any she could see possible. When I asked what her fears for the future were, she named a litany: facing more pain, suffering the humiliation of losing more of her bodily control, being unable to leave the hospital. She choked up as she spoke. She'd been there for days just getting worse, and she feared she didn't have many more. I asked her if they'd talked to her about hospice. They had, she said, but she didn't see what it could do to help her.

Some in her position, offered "death with dignity," might have taken it as the only chance for control when no other options seemed apparent. Martin and I persuaded Peg to try hospice. It'd at least let her get home, I said, and might help her more than she knew. I explained how hospice's aim, at least in theory, was to give people their best possible day, however they might define it under the circumstances. It seemed like it had been a while since she'd had a good day, I said.

"Yes, it has—a long while," she said.

That seemed worth hoping for, I said—just one good day.

She went home on hospice within forty-eight hours. We broke the news to Hunter that Peg would not be able to give her lessons anymore, that she was dying. Hunter was struck low. She adored Peg. She wanted to know if she could see her one more time. We had to tell her that we didn't think so.

A few days later, we got a surprising call. It was Peg. If Hunter was willing, she said, she'd like to resume teaching her. She'd understand if Hunter didn't want to come. She didn't know how many more lessons she could manage, but she wanted to try.

That hospice could make it possible for her to teach again was more than I'd ever imagined, certainly more than she'd imagined. But when her hospice nurse, Deborah, arrived, they began talking about what Peg cared most about in her life, what having the best day possible would really mean to her. Then they worked together to make it happen.

At first, her goal was just managing her daily difficulties. The hospice team set up a hospital bed on the first floor so she wouldn't have to navigate the stairs. They put a portable commode at the bedside. They organized help for bathing and getting dressed. They gave her morphine, gabapentin, and oxycodone to control her pain, and methylphenidate proved helpful for combating the stupor they induced.

Her anxieties plummeted as the challenges came under control. She raised her sights. "She was focused on the main chance," Martin later said. "She came to a clear view of how she wanted to live the rest of her days. She was going to be home, and she was going to teach."

It took planning and great expertise to make each lesson possible. Deborah helped her learn how to calibrate her medications. "Before she would teach, she would take some additional morphine. The trick was to give her enough to be comfortable to teach and not so much that she would be groggy," Martin recalled.

Nonetheless, he said, "She was more alive running up to a lesson and for the days after." She'd had no children; her students filled that place for her. And she still had some things she wanted them to know before she went. "It was important to her to be able to say her good-byes to her dear friends, to give her parting advice to her students."

She lived six full weeks after going on hospice. Hunter had lessons for four of them, and then two final concerts were played. One featured Peg's former students, accomplished performers from around the country, the other her current students, all children in middle school and high school. Gathered together in her living room, they played Brahms, Dvořák, Chopin, and Beethoven for their adored teacher.

Technological society has forgotten what scholars call the "dying role" and its importance to people as life approaches its end. People want to share memories, pass on wisdoms and keepsakes, settle relationships, establish their legacies, make peace with God, and ensure that those who are left behind will be okay. They want to end their stories on their own terms. This role is, observers argue, among life's most important, for both the dying and those left behind. And if it is, the way we deny people this role, out of obtuseness and neglect, is cause for everlasting shame. Over and over, we in medicine inflict deep gouges at the end of people's lives and then stand oblivious to the harm done.

Peg got to fulfill her dying role. She got to do so right up to three days before the end, when she fell into delirium and passed in and out of consciousness.

My final remembrance of her is from near the end of her last recital. She'd taken Hunter away from the crowd and given her a book of music she wanted her to keep. Then she put her arm around her shoulder.

"You're special," she whispered to her. It was something she never wanted Hunter to forget.

EVENTUALLY, THE TIME came for my father's story to end, as well. For all our preparations and all I thought I had learned, we weren't ready for it, though. Ever since he'd gotten on hospice in the early spring, he'd arrived at what seemed like a new, imperfect, but manageable steady state. Between my mother, the various helpers she had arranged, and his own steel will, he'd been able to string together weeks of good days.

Each had its sufferings and humiliations, to be sure. He needed daily enemas. He soiled the bed. The pain medications made his head feel "fuzzy," "foggy," "heavy," he said, and he disliked that intensely. He did not want to be sedated; he wanted to be able to see people and communicate. Pain, however, was far worse. If he lightened up on the dose of his medications, he experienced severe headaches and a lancing pain that shot up and down his neck and back. When he was in the grip of it, the pain became his entire world. He tinkered constantly with his doses, trying to find the combination that would let him feel neither pain nor fogginess—feel normal, like the person he'd been before his body began failing him. But no matter what the drug or dose, normal was out of reach.

Good enough, however, could be found. Through the spring and early summer, he still had dinner parties at which he'd preside from the head of the table. He made plans for a new building at the college in India. He sent out a dozen e-mails a day, despite the difficulty controlling his weakened hands. He and my mother watched a movie together almost every night and cheered on Novak Djokovic through his two-week run to victory at Wimbledon. My sister brought home her new boyfriend, whom she felt might be "the one"—they did in fact eventually marry—and my father was bowled over with happiness for her. Each day, he found moments worth living for. And as the weeks stretched into months, it seemed like he could continue this way a long time.

In retrospect, there were signs that he couldn't. His weight continued to drop. The doses of pain medication he required were increasing. During the first couple days of August, I received a series of garbled e-mails. "Dear Atuli whohirnd li9ke Sude," began one. The last one said:

Dear Atul
sorry for scrambeled letth ter. i having problems.
-With love
Dad-

On the phone, he spoke more slowly, with long pauses between sentences. He explained that he sometimes felt confused and was having trouble communicating. His e-mails were not making sense to him, he said, although he thought they did when he first wrote them. His world was closing in.

Then on Saturday, August 6, at 8:00 a.m., my mother called, frightened. "He's not waking up," she said. He was breathing, but she couldn't rouse him. It was the medication, we thought. The night before he'd insisted on taking a whole tablet of buprenorphine, a narcotic pill, instead of a half pill like he'd been taking, my mother explained. She'd argued with him, but he'd become angry. He wanted no pain, he said. Now he wasn't waking up. In doctor mode, she noted his pinpoint pupils, a sign of a narcotic overdose. We decided to wait it out and let the medication wear off.

Three hours later, she phoned again. She had called an ambulance, not the hospice agency. "He was turning blue, Atul." She was in the hospital emergency room. "His blood pressure is fifty. He's still not waking up. His oxygen is low." The medical staff gave him naloxone, a narcotic-reversal agent, and if he had overdosed, that should have woken him. But he remained unresponsive. A stat chest X-ray showed pneumonia in his right lung. They gave him a face mask with 100 percent oxygen, antibiotics, and fluids. But his oxygen level would not come up above 70 percent, an unsurvivable level. Now, my mother said, they were asking whether they should intubate him, put him on drips to support his blood pressure, and move him to the ICU. She didn't know what to do.

As a person's end draws near, there comes a moment when responsibility shifts to someone else to decide what to do. And we'd mostly prepared for that moment. We'd had the hard conversations. He'd already spelled out how he wanted the end of his story to be written. He wanted no ventilators and no suffering. He wanted to remain home and with the people he loved.

But the arrow of events refuses to follow a steady course and that plays havoc with a surrogate's mind. Only the day before, it seemed he might have weeks, even months. Now she was supposed to believe that hours might be a stretch? My mother's heart was breaking, but as we talked, she recognized the pathway we risked heading down, and that the kind of life intensive care would preserve for him was far from the one he wanted. Endings matter, not just for the person but, perhaps even more, for the ones left behind. She decided to tell them not to intubate him. I called my sister and caught her as she was about to board her train into work. She was not ready for the news, either.

"How could it be?" she asked. "Are we certain he can't return to how he was yesterday?"

"It seems unlikely," I said. In few families does everyone see such situations the same. I arrived the quickest at the idea that my father was coming to the end, and I worried most about the mistake of prolonging his suffering too long. I saw the opportunity for a peaceful end as a blessing. But to my sister, and even more my mother, it didn't seem certain at all that he was at the end, and the mistake that loomed largest to them was the possibility of failing to preserve his life long enough. We agreed not to let the hospital do anything further to resuscitate him, while hoping against hope that he'd hang on long enough for my sister and me to get there to see him. We both searched for flights as they moved him to a private hospital room.

Later that afternoon, my mother called as I sat at an airport departure gate.

"He's awake!" she said, over the moon. He'd recognized her. He was sharp enough to even ask what his blood pressure was. I felt abashed for believing that he wouldn't come to. No matter how much one has seen, nature refuses predictability. More than this, though, what I kept thinking was: I'm going to be there. He may even be all right for a while longer.

He was alive just four more days, as it turned out. When I arrived at his bedside, I found him alert and unhappy about awaking in the hospital. No one listens to him, he said. He'd awoken in severe pain but the medical staff wouldn't give him enough medication to stop it, fearing he might lose consciousness again. I asked the nurse to give him the full dose he took at home. She had to get permission from the doctor on call, and still he approved only half.

Finally, at 3:00 a.m., my father had had enough. He began shouting. He demanded that they take out his IVs and let him go home. "Why are you doing nothing?" he yelled. "Why are you letting me suffer?" He'd become incoherent with pain. He called the Cleveland Clinic—two hundred miles away—on his cell phone and told a confused doctor on duty to "Do something." His night nurse finally got permission for a slug of an intravenous narcotic, but he refused it. "It doesn't work," he said. Finally, at 5:00 a.m., we persuaded him to take the injection, and the pain began to subside. He became calm. But he still wanted to go home. In a hospital built to ensure survival at all costs and unclear how to do otherwise, he understood his choices would never be his own.

We arranged for the medical staff to give him his morning

dose of medication, stop his oxygen and his antibiotics for his pneumonia, and let us take him. By midmorning he was back in his bed.

"I do not want suffering," he repeated when he had me alone. "Whatever happens, will you promise me you won't let me suffer?"

"Yes," I said.

That was harder to achieve than it would seem. Just urinating, for instance, proved a problem. His paralysis had advanced from just the week before, and one sign was that he became unable to pee. He could still feel when his bladder became full but could make nothing come out. I helped him to the bathroom and swiveled him onto the seat. Then I waited while he sat there. Half an hour passed. "It'll come," he insisted. He tried not to think about it. He pointed out the toilet seat from Lowe's he'd had installed a couple months before. It was electric, he said. He loved it. It could wash his bottom with a burst of water and dry it. No one had to wipe him. He could take care of himself.

"Have you tried it?" he asked.

"That would be no," I said.

"You ought to," he said, smiling.

But still nothing came out. Then the bladder spasms began. He groaned when they came over him. "You're going to have to catheterize me," he said. The hospice nurse, expecting this moment would come, had brought the supplies and trained my mother. But I'd done it a hundred times for my own patients. So I pulled my father up from the seat, got him back to bed, and set about doing it for him, his eyes squeezed shut the entire time. It's not something a person ever thinks they will come to. But I got the catheter in, and the urine flooded out. The relief was oceanic.

His greatest struggle remained the pain from his tumor—

not because it was difficult to control but because it was difficult to agree on how much to control it. By the third day, he'd become unarousable again for long periods. The question became whether to keep giving him his regular dose of liquid morphine, which could be put under his tongue where it would absorb into his bloodstream through his mucous membranes. My sister and I thought we should, fearing that he might wake up in pain. My mother thought we shouldn't, fearing the opposite.

"Maybe if he had a little pain, he'd wake up," she said, her eyes welling. "He still has so much he can do."

Even in his last couple of days, she was not wrong. When he was permitted to rise above the demands of his body, he took the opportunity for small pleasures greedily. He could still enjoy certain foods and ate surprisingly well, asking for chapatis, rice, curried string beans, potatoes, yellow split-pea dahl, black-eyed-pea chutney, and *shira*, a sweet dish from his youth. He talked to his grandchildren by phone. He sorted photos. He gave instructions about unfinished projects. He had but the tiniest fragments of life left that he could grab, and we were agonizing over them. Could we get him another one?

Nonetheless, I remembered my pledge to him and gave him his morphine every two hours, as planned. My mother anxiously accepted it. For long hours, he lay quiet and stock-still, except for the rattle of his breathing. He'd have a sharp intake of breath—it sounded like a snore that would shut off suddenly, as if a lid had come down—followed a second later by a long exhale. The air rushing past the mucoid fluid in his windpipe sounded like someone shaking pebbles in a hollow tube in his chest. Then there'd be silence for what seemed like forever before the cycle would start up again.

We got used to it. He lay with his hands across his belly, peaceful, serene. We sat by his bedside for long hours, my mother reading the *Athens Messenger*, drinking tea, and worrying whether my sister and I were getting enough to eat. It was comforting to be there.

Late on his penultimate afternoon, he broke out into a soaking sweat. My sister suggested that we change his shirt and wash him. We lifted him forward, into a sitting position. He was unconscious, a completely dead weight. We tried getting his shirt over his head. It was awkward work. I tried to remember how nurses do it. Suddenly I realized his eyes were open.

"Hi, Dad," I said. He just looked for a while, observing, breathing hard.

"Hi," he said.

He watched as we cleaned his body with a wet cloth, gave him a new shirt.

"Do you have any pain?"

"No." He motioned that he wanted to get up. We got him into a wheelchair and took him to a window looking out onto the backyard, where there were flowers, trees, sun on a beautiful summer day. I could see that his mind was gradually clearing.

Later, we wheeled him to the dinner table. He had some mango, papaya, yogurt, and his medications. He was silent, breathing normally again, thinking.

"What are you thinking?" I asked.

"I'm thinking how to not prolong the process of dying. This—this food prolongs the process."

My mom didn't like hearing this.

"We're happy taking care of you, Ram," she said. "We love you."

He shook his head.

"It's hard, isn't it?" my sister said.

"Yes. It's hard."

"If you could sleep through it, is that what you'd prefer?" I asked.

"Yes."

"You don't want to be awake, aware of us, with us like this?" my mother asked.

He didn't say anything for a moment. We waited.

"I don't want to experience this," he said.

The suffering my father experienced in his final day was not exactly physical. The medicine did a good job of preventing pain. When he surfaced periodically, at the tide of consciousness, he would smile at our voices. But then he'd be fully ashore and realize that it was not over. He'd realize that all the anxieties of enduring that he'd hoped would be gone were still there: the problems with his body, yes, but more difficult for him the problems with his mind—the confusion, the worries about his unfinished work, about Mom, about how he'd be remembered. He was at peace in sleep, not in wakefulness. And what he wanted for the final lines of his story, now that nature was pressing its limits, was peacefulness.

During his final bout of wakefulness, he asked for the grandchildren. They were not there, so I showed him pictures on my iPad. His eyes went wide, and his smile was huge. He looked at every picture in detail.

Then he descended back into unconsciousness. His breathing stopped for twenty or thirty seconds at a time. I'd be sure it was over, only to find that his breathing would start again. It went on this way for hours.

Finally, around ten after six in the afternoon, while my mother and sister were talking and I was reading a book, I noticed that he'd stopped breathing for longer than before.

"I think he's stopped," I said.

We went to him. My mother took his hand. And we listened, each of us silent.

No more breaths came.

Epilogue

Being mortal is about the struggle to cope with the constraints of our biology, with the limits set by genes and cells and flesh and bone. Medical science has given us remarkable power to push against these limits, and the potential value of this power was a central reason I became a doctor. But again and again, I have seen the damage we in medicine do when we fail to acknowledge that such power is finite and always will be.

We've been wrong about what our job is in medicine. We think our job is to ensure health and survival. But really it is larger than that. It is to enable well-being. And well-being is about the reasons one wishes to be alive. Those reasons matter not just at the end of life, or when debility comes, but all along the way. Whenever serious sickness or injury strikes and your body or mind breaks down, the vital questions are the same: What is your understanding of the situation and its potential outcomes? What are your fears and what are your hopes? What are the trade-offs you are willing to make and not willing to make? And what is the course of action that best serves this understanding?

The field of palliative care emerged over recent decades to

bring this kind of thinking to the care of dying patients. And the specialty is advancing, bringing the same approach to other seriously ill patients, whether dying or not. This is cause for encouragement. But it is not cause for celebration. That will be warranted only when all clinicians apply such thinking to every person they touch. No separate specialty required.

If to be human is to be limited, then the role of caring professions and institutions—from surgeons to nursing homes—ought to be aiding people in their struggle with those limits. Sometimes we can offer a cure, sometimes only a salve, sometimes not even that. But whatever we can offer, our interventions, and the risks and sacrifices they entail, are justified only if they serve the larger aims of a person's life. When we forget that, the suffering we inflict can be barbaric. When we remember it the good we do can be breathtaking.

I never expected that among the most meaningful experiences I'd have as a doctor—and, really, as a human being—would come from helping others deal with what medicine cannot do as well as what it can. But it's proved true, whether with a patient like Jewel Douglass, a friend like Peg Bachelder, or someone I loved as much as my father.

MY FATHER CAME to his end never having to sacrifice his loyalties or who he was, and for that I am grateful. He was clear about his wishes even for after his death. He left instructions for my mother, my sister, and me. He wanted us to cremate his body and spread the ashes in three places that were important to him—in Athens, in the village where he'd grown up, and on the Ganges River, which is sacred to all Hindus. According to Hindu mythology, when a person's remains touch the great river, he or she is assured eternal salvation. So for millennia, families have brought the ashes of their loved ones to the Ganges and spread them upon its waters.

A few months after my father's death we therefore followed in those footsteps. We traveled to Varanasi, the ancient city of temples on the banks of the Ganges, which dates back to the twelfth century BC. Waking before the sun rose, we walked out onto the ghats, the walls of steep steps lining the banks of the massive river. We'd secured ahead of time the services of a pandit, a holy man, and he guided us onto a small wooden boat with a rower who pulled us out onto the predawn river.

The air was crisp and chilly. A shroud of white fog hung over the city's spires and the water. A temple guru sang mantras broadcast over staticky speakers . The sound drifted across the river to the early bathers with their bars of soap, the rows of washermen beating clothes on stone tablets, and a kingfisher sitting on a mooring. We passed riverbank platforms with huge stacks of wood awaiting the dozens of bodies to be cremated that day. When we'd traveled far enough out into the river and the rising sun became visible through the mist, the pandit began to chant and sing.

As the oldest male in the family, I was called upon to assist with the rituals required for my father to achieve *moksha*— liberation from the endless earthly cycle of death and rebirth to ascend to nirvana. The pandit twisted a ring of twine onto the fourth finger of my right hand. He had me hold the palm-size brass urn that contained my father's ashes and sprinkle into it herbal medicines, flowers, and morsels of food: a betel nut, rice, currants, rock crystal sugar, turmeric. He then had the other members of the family do the same. We burned incense and wafted the smoke over the ashes. The pandit reached over the bow with a small cup and had me drink three tiny spoons of Ganga water. Then he told me to throw the urn's dusty contents over my right shoulder into the river, followed by the urn itself and its cap. "Don't look," he admonished me in English, and I didn't.

It's hard to raise a good Hindu in small-town Ohio, no matter how much my parents tried. I was not much of a believer in the idea of gods controlling people's fates and did not suppose that anything we were doing was going to offer my father a special place in any afterworld. The Ganges might have been sacred to one of the world's largest religions, but to me, the doctor, it was more notable as one of the world's most polluted rivers, thanks in part to all the incompletely cremated bodies that had been thrown into it. Knowing that I'd have to take those little sips of river water, I had looked up the bacterial counts on a Web site beforehand and premedicated myself with the appropriate antibiotics. (Even so, I developed a *Giardia* infection, having forgotten to consider the possibility of parasites.)

Yet I was still intensely moved and grateful to have gotten to do my part. For one, my father had wanted it, and my mother and sister did, too. Moreover, although I didn't feel my dad was anywhere in that cup and a half of gray, powdery ash, I felt that we'd connected him to something far bigger than ourselves, in this place where people had been performing these rituals for so long.

When I was a child, the lessons my father taught me had been about perseverance: never to accept limitations that stood in my way. As an adult watching him in his final years, I also saw how to come to terms with limits that couldn't simply be wished away. When to shift from pushing against limits to making the best of them is not often readily apparent. But it is clear that there are times when the cost of pushing exceeds its value. Helping my father through the struggle to define that moment was simultaneously among the most painful and most privileged experiences of my life.

Part of the way my father handled the limits he faced was by looking at them without illusion. Though his circumstances sometimes got him down, he never pretended they were better than they were. He always understood that life is short and one's place in the world is small. But he also saw himself as a link in a chain of history. Floating on that swollen river, I could not help sensing the hands of the many generations connected across time. In bringing us there, my father had helped us see that he was part of a story going back thousands of years—and so were we.

We were lucky to get to hear him tell us his wishes and say his good-byes. In having a chance to do so, he let us know he was at peace. That let us be at peace, too.

After spreading my father's ashes, we floated silently for a while, letting the current take us. As the sun burned away the mist, it began warming our bones. Then we gave a signal to the boatman, and he picked up his oars. We headed back toward the shore.

Notes on Sources

INTRODUCTION

1   Tolstoy's classic novella: Leo Tolstoy, *The Death of Ivan Ilyich*, 1886 (Signet Classic, 1994).

3   I began writing: A. Gawande, *Complications* (Metropolitan Books, 2002).

6   As recently as 1945: National Office of Vital Statistics, *Vital Statistics of the United States, 1945* (Government Printing Office, 1947), p. 104, http://www.cdc.gov/nchs/data/vsus/vsus_1945_1.pdf.

6   In the 1980s: J. Flory et al., "Place of Death: U.S. Trends since 1980," *Health Affairs* 23 (2004): 194–200, http://content.healthaffairs.org /content/23/3/194.full.html.

6   Across not just the United States: A. Kellehear, *A Social History of Dying* (Cambridge University Press, 2007).

8   The late surgeon Sherwin Nuland: S. Nuland, *How We Die: Reflections on Life's Final Chapter* (Knopf, 1993).

1: THE INDEPENDENT SELF

17   Even when the nuclear family: P. Thane, ed., *A History of Old Age* (John Paul Getty Museum Press, 2005).

17   one child usually remained: D. H. Fischer, *Growing Old in America: The Bland-Lee Lectures Delivered at Clark University* (Oxford University Press, 1978). Also C. Haber and B. Gratton, *Old Age and the*

Search for Security: An American Social History (Indiana University Press, 1994).

17   the poet Emily Dickinson: C. A. Kirk, *Emily Dickinson: A Biography* (Greenwood Press, 2004).

18   surviving into old age was uncommon: R. Posner, *Aging and Old Age* (University of Chicago Press, 1995), see ch. 9.

18   They tended to maintain their status . . . Whereas today people often understate: Fischer, *Growing Old in America*.

18   In America, in 1790: A. Achenbaum, *Old Age in the New Land* (Johns Hopkins University Press, 1979).

18   today, they are 14 percent: United States Census Bureau, http://quick facts.census.gov/qfd/states/00000.html.

18   In Germany, Italy, and Japan: World Bank, http://data.worldbank.org /indicator/SP.POP.65UP.TO.ZS.

18   100 million elderly: "China's Demographic Time Bomb," *Time*, Aug. 31, 2011, http://www.time.com/time/world/article/0,8599,2091308, 00.html.

18   As for the exclusive hold: Posner, ch. 9.

18   increased longevity has brought: Haber and Gratton, pp. 24–25, 39.

20   Historians find that the elderly . . . The radical concept of "retirement": Haber and Gratton.

21   Life expectancy: E. Arias, "United States Life Tables," *National Vital Statistics Reports* 62 (2014): 51.

21   Family sizes fell: L. E. Jones and M. Tertilt, "An Economic History of Fertility in the U.S., 1826–1960," *NBER Working Paper Series*, Working Paper 12796, 2006, http://www.nber.org/papers/w12796.

21   The average age at which: Fischer, appendix, table 6.

21   "intimacy at a distance": L. Rosenmayr and E. Kockeis, "Propositions for a Sociological Theory of Aging and the Family," *International Social Science Journal* 15 (1963): 410–24.

21   Whereas in early-twentieth-century America: Haber and Gratton, p. 44.

21   The pattern is a worldwide one: E. Klinenberg, *Going Solo: The Extraordinary Rise and Surprising Appeal of Living Alone* (Penguin, 2012).

21   Just 10 percent: European Commission, *i2010: Independent Living for the Ageing Society*, http://ec.europa.eu/information_society/activ ities/ict_psp/documents/independent_living.pdf.

21   Del Webb: J. A. Trolander, *From Sun Cities to the Villages* (University Press of Florida, 2011).

25   trajectory of our lives: J. R. Lunney et al., "Patterns of Functional
      Decline at the End of Life," *Journal of the American Medical Asso-*
      ciation 289 (2003): 2387–92. The graphs in this chapter are adapted
      from the article.

26   By the middle of the twentieth century: National Center for Health
      Statistics, *Health, United States, 2012: With Special Feature on*
      Emergency Care (Washington, DC: U.S. Government Printing Office,
      2013).

26–28 People with incurable cancers ... The curve of life becomes a long
      slow fade: J. R. Lunney, J. Lynn, and C. Hogan, "Profiles of Older
      Medicare Decedents," *Journal of the American Geriatrics Society*
      50 (2002): 1109. See also Lunney et al., "Patterns of Functional
      Decline."

29   Consider the teeth: G. Gibson and L. C. Niessen, "Aging and the
      Oral Cavity," in *Geriatric Medicine: An Evidence-Based Approach*,
      ed. C. K. Cassel (Springer, 2003), pp. 901–19. See also I. Barnes and
      A. Walls, "Aging of the Mouth and Teeth," *Gerodontology* (John
      Wright, 1994).

29   the muscles of the jaw lose: J. R. Drummond, J. P. Newton, and
      R. Yemm, *Color Atlas and Text of Dental Care of the Elderly*
      (Mosby-Wolfe, 1995), pp. 49–50.

29   By the age of sixty: J. J. Warren et al., "Tooth Loss in the Very Old:
      13-15-Year Incidence among Elderly Iowans," *Community Den-*
      tistry and Oral Epidemiology 30 (2002): 29–37.

30   Under a microscope: A. Hak et al., "Progression of Aortic Calcifi-
      cation Is Associated with Metacarpal Bone Loss during Meno-
      pause: A Population-Based Longitudinal Study," *Arteriosclerosis,*
      Thrombosis, and Vascular Biology 20 (2000): 1926–31.

30   Research has found that loss of bone density: H. Yoon et al., "Cal-
      cium Begets Calcium: Progression of Coronary Artery Calcification
      in Asymptomatic Subjects," *Radiology* 224 (2002): 236–41; Hak et
      al., "Progression of Aortic Calcification."

30   more than half of us: N. K. Wenger, "Cardiovascular Disease," in
      Geriatric Medicine, ed. Cassel (Springer, 2003); B. Lernfeit et al.,
      "Aging and Left Ventricular Function in Elderly Healthy People,"
      American Journal of Cardiology 68 (1991): 547–49.

30   muscle elsewhere thins: J. D. Walston, "Sarcopenia in Older Adults,"

Current Opinion in Rheumatology 24 (2012): 623–27; E. J. Metter et al., "Age-Associated Loss of Power and Strength in the Upper Extremities in Women and Men," *Journal of Gerontology: Biological Sciences* 52A (1997): B270.

30 You can see all these processes: E. Carmeli, "The Aging Hand," *Journal of Gerontology: Medical Sciences* 58A (2003): 146–52.

31 This is normal: R. Arking, *The Biology of Aging: Observations and Principles*, 3rd ed. (Oxford University Press, 2006); A. S. Dekaban, "Changes in Brain Weights During the Span of Human Life: Relation of Brain Weights to Body Heights and Body Weights," *Annals of Neurology* 4 (1978): 355; R. Peters, "Ageing and the Brain," *Postgraduate Medical Journal* 82 (2006): 84–85; G. I. M. Craik and E. Bialystok, "Cognition Through the Lifespan: Mechanisms of Change," *Trends in Cognitive Sciences* 10 (2006): 132; R. S. N. Liu et al., "A Longitudinal Study of Brain Morphometrics Using Quantitative Magentic Resonance Imaging and Difference Image Analysis," *NeuroImage* 20 (2003): 26; T. A. Salthouse, "Aging and Measures of Processing Speed," *Biological Psychology* 54 (2000): 37; D. A. Evans et al., "Prevalence of Alzheimer's Disease in a Community Population of Older Persons," *JAMA* 262 (1989): 2251.

31 Why we age: R. E. Ricklefs, "Evolutionary Theories of Aging: Confirmation of a Fundamental Prediction, with Implications for the Genetic Basis and Evolution of Life Span," *American Naturalist* 152 (1998): 24–44; R. M. Zammuto, "Life Histories of Birds: Clutch Size, Longevity, and Body Mass among North American Game Birds," *Canadian Journal of Zoology* 64 (1986): 2739–49.

32 The idea that living things shut down: C. Mobbs, "Molecular and Biologic Factors in Aging," in *Geriatric Medicine*, ed. Cassel; L. A. Gavrilov and N. S. Gavrilova, "Evolutionary Theories of Aging and Longevity," *Scientific World Journal* 2 (2002): 346.

32 average life span of human beings: S. J. Olshansky, "The Demography of Aging," in *Geriatric Medicine*, ed. Cassel; Kellehear, *A Social History*.

32 As Montaigne wrote: Michel de Montaigne. *The Essays*, sel. and ed. Adolphe Cohn (G. P. Putnam's Sons, 1907), p. 278.

33 inheritance has surprisingly little influence: G. Kolata, "Live Long? Die Young? Answer Isn't Just in Genes," *New York Times*, Aug. 31, 2006; K. Christensen and A. M. Herskind, "Genetic Factors Associated with Individual Life Duration: Heritability," in J. M. Robine et al., eds., *Human Longevity, Individual Life Duration, and the Growth of the Oldest-Old Population* (Springer, 2007).

33 If our genes explain less: Gavrilov and Gavrilova, "Evolutionary Theories of Aging and Longevity."

34 Hair grows gray: A. K. Freeman and M. Gordon, "Dermatologic Diseases and Problems," in *Geriatric Medicine*, ed. Cassel, 869.

34 Inside skin cells: A. Terman and U. T. Brunk, "Lipofuscin," *International Journal of Biochemistry and Cell Biology* 36 (2004): 1400–4; Freeman and Gordon, "Dermatologic Diseases and Problems."

34 The eyes go: R. A. Weale, "Age and the Transmittance of the Human Crystalline Lens," *Journal of Physiology* 395 (1988): 577–87.

35 the "rectangularization" of survival: Olshansky, "The Demography of Aging." See also US Census Bureau data for 1950, http://www.census.gov/ipc/www/idbpyr.html. Additional data from Population Pyramid online, http://populationpyramid.net/.

36 We cling to the notion of retirement: M. E. Pollack, "Intelligent Technology for an Aging Population: The Use of AI to Assist Elders with Cognitive Impairment," *AI Magazine* (Summer 2005): 9–25. See also Federal Deposit Insurance Corporation, *Economic Conditions and Emerging Risks in Banking: A Report to the FDIC Board of Directors*, May 9, 2006, http://www.fdic.gov/deposit/insurance/risk/2006_02/Economic_2006_02.html.

36 Equally worrying: Data on certifications in geriatrics from American Board of Medical Specialties and American Board of Internal Medicine.

40 350,000 Americans fall and break a hip: M. Gillick, *The Denial of Aging: Perpetual Youth, Eternal Life, and Other Dangerous Fantasies* (Harvard University Press, 2006).

44 Several years ago, researchers at the University of Minnesota: C. Boult et al., "A Randomized Clinical Trial of Outpatient Geriatric Evaluation and Management," *Journal of the American Geriatrics Society* 49 (2001): 351–59.

52 In a year, fewer than three hundred doctors: American Board of Medical Specialties, American Board of Psychiatry and Neurology; L. E. Garcez-Leme et al., "Geriatrics in Brazil: A Big Country with Big Opportunities," *Journal of the American Geriatrics Society* 53 (2005): 2018–22; C. L. Dotchin et al., "Geriatric Medicine: Services and Training in Africa," *Age and Ageing* 41 (2013): 124–28.

53 The risk of a fatal car crash: D. C. Grabowski, C. M. Campbell, and M. A. Morrissey, "Elderly Licensure Laws and Motor Vehicle Fatalities," *JAMA* 291 (2004): 2840–46.

53 in Los Angeles, George Weller: J. Spano, "Jury Told Weller Must Pay

for Killing 10," *Los Angeles Times*, Oct. 6, 2006, http://articles.la
times.com/2006/oct/06/local/me-weller6.

3: DEPENDENCE

61 In 1913, Mabel Nassau: M. L. Nassau, *Old Age Poverty in Greenwich Village: A Neighborhood Study* (Fleming H. Revell Co., 1915).

62 Unless family could take such people in: M. Katz, *In the Shadow of
the Poorhouse* (Basic Books, 1986); M. Holstein and T. R. Cole, "The
Evolution of Long-Term Care in America," in *The Future of Long-
Term Care*, ed. R. H. Binstock, L. E. Cluff, and O. Von Mering (Johns
Hopkins University Press, 1996).

62 A 1912 report: Illinois State Charities Commission, *Second Annual
Report of the State Charities Commission*, 1912, pp. 457–508; Virginia State Board of Charities and Corrections, *First Annual Report
of State Board of Charities and Corrections*, 1909.

63 Nothing provoked greater terror: Haber and Gratton, *Old Age and
the Search for Security*.

66 the case of Harry Truman: M. Barber, "Crotchety Harry Truman
Remains an Icon of the Eruption," *Seattle Post-Intelligencer*, March 11,
2000; S. Rosen, *Truman of Mt. St. Helens: The Man and His Mountain* (Madrona Publishers, 1981). Two bands have put out songs
inspired by Truman: R. W. Stone's 1980 country rock hit, "Harry Truman, Your Spirit Still Lives On," http://www.youtube.com/watch
?v=WGwa3N43GB4, and Headgear's 2007 indie rock single, "Harry
Truman," http://www.youtube.com/watch?v=JvcZnKkM_DE.

69 In the middle part of the twentieth century: L. Thomas, *The Youngest
Science* (Viking, 1983).

69 Congress passed the Hill-Burton Act: A. P. Chung, M. Gaynor, and S.
Richards-Shubik, "Subsidies and Structure: The Last Impact of the
Hill-Burton Program on the Hospital Industry," National Bureau of
Economics Research Program on Health Economics meeting paper,
April 2013, http://www.nber.org/confer/2013/HEs13/summary.htm.

70 Meanwhile, policy planners: A key source for the history of nursing
homes was B. Vladeck, *Unloving Care: The Nursing Home Tragedy*
(Basic Books, 1980). See also Holstein and Cole, "Evolution of Long-
Term Care," and records from the City of Boston and its almshouse:
https://www.cityofboston.gov/Images_Documents/Guide%20to
%20the%20Almshouse%20records_tcm3-30021.pdf.

71 As one scholar put it: Vladeck, *Unloving Care*.

73   The sociologist Erving Goffman: E. Goffman *Asylums* (Anchor, 1961). Corroborated by C. W. Lidz, L. Fischer, and R. M. Arnold, *The Erosion of Autonomy in Long-Term Care* (Oxford University Press, 1992).

4: ASSISTANCE

79   Your chances of avoiding the nursing home: G. Spitze and J. Logan, "Sons, Daughters, and Intergenerational Social Support," *Journal of Marriage and Family* 52 (1990): 420–30.

88   "Her vision was simple": K. B. Wilson, "Historical Evolution of Assisted Living in the United States, 1979 to the Present," *Gerontologist* 47, special issue 3 (2007): 8–22.

92   In 1988, the findings were made public: K. B. Wilson, R. C. Ladd, and M. Saslow, "Community Based Care in an Institution: New Approaches and Definitions of Long Term Care" paper presented at the 41st Annual Scientific Meeting of the Gerontological Society of America, San Francisco, Nov. 1988. Cited in Wilson, "Historical Evolution."

92–93 In 1943, the psychologist Abraham Maslow: A. H. Maslow, "A Theory of Human Motivation," *Psychological Review* 50 (1943): 370–96.

94   Studies find that as people grow older: D. Field and M. Minkler, "Continuity and Change in Social Support between Young-Old, Old-Old, and Very-Old adults," *Journal of Gerontology* 43 (1988): 100–6; K. Fingerman and M. Perlmutter, "Future Time Perspective and Life Events across Adulthood," *Journal of General Psychology* 122 (1995): 95–111.

94   In one of her most influential studies: L. L. Carstensen et al., "Emotional Experience Improves with Age: Evidence Based on over 10 Years of Experience Sampling," *Psychology and Aging* 26 (2011): 21–33.

98   She produced a series of experiments: L. L. Carstensen and B. L. Fredrickson, "Influence of HIV Status on Cognitive Representation of Others," *Health Psychology* 17 (1998): 494–503; H. H. Fung, L. L. Carstensen, and A. Lutz, "Influence of Time on Social Preferences: Implications for Life-Span Development," *Psychology and Aging* 14 (1999): 595; B. L. Fredrickson and L. L. Carstensen, "Choosing Social Partners: How Old Age and Anticipated Endings Make People More Selective," *Psychology and Aging* 5 (1990): 335; H. H. Fung and L. L. Carstensen, "Goals Change When Life's Fragility Is Primed: Lessons Learned from Older Adults, the September 11 Attacks, and SARS," *Social Cognition* 24 (2006): 248–78.

101   By 2010, the number of people in assisted living: Center for Medicare and Medicaid Services, *Nursing Home Data Compendium, 2012 Edition* (Government Printing Office, 2012).

102   A survey of fifteen hundred assisted living facilities: C. Hawes et al., "A National Survey of Assisted Living Facilities," *Gerontologist* 43 (2003): 875–82.

5: A BETTER LIFE

122   In a book he wrote: W. Thomas, *A Life Worth Living* (Vanderwyk and Burnham, 1996).

123–24   And other research was consistent: J. Rodin and E. Langer, "Long-Term Effects of a Control-Relevant Intervention with the Institutionalized Aged," *Journal of Personality and Social Psychology* 35 (1977): 897–902.

125   In 1908, a Harvard philosopher: J. Royce, *The Philosophy of Loyalty* (Macmillan, 1908).

129   Research has found that in units with fewer than twenty people: M. P. Calkins, "Powell Lawton's Contributions to Long-Term Care Settings," *Journal of Housing for the Elderly* 17 (2008): 1–2, 67–84.

140   As Dworkin wrote: R. Dworkin, "Autonomy and the Demented Self," *Milbank Quarterly* 64, supp. 2 (1986): 4–16.

6: LETTING GO

150   More than 15 percent of lung cancers: C. M. Rudin et al., "Lung Cancer in Never Smokers: A Call to Action," *Clinical Cancer Research* 15 (2009): 5622–25.

151   85 percent of them respond: C. Zhou et al., "Erlotinib versus Chemotherapy for Patients with Advanced EGFR Mutation-Positive Non-Small-Cell Lung Cancer," *Lancet Oncology* 12 (2011): 735–42.

152   Studies had shown: C. P. Belani et al., "Maintenance Pemetrexed plus Best Supportive Care (BSC) versus Placebo plus BSC: A Randomized Phase III Study in Advanced Non-Small Cell Lung Cancer," *Journal of Clinical Oncology* 27 (2009): 18s.

153   In the United States, 25 percent of all Medicare spending: G. F. Riley and J. D. Lubitz, "Long-Term Trends in Medicare Payments in the Last Year of Life," *Health Services Research* 45 (2010): 565–76.

153 Data from elsewhere: L. R. Shugarman, S. L. Decker, and A. Bercovitz, "Demographic and Social Characteristics and Spending at the End of Life," *Journal of Pain and Symptom Management* 38 (2009): 15–26.

153 Spending on a disease like cancer: A. B. Mariotto, K. R. Yabroff, Y. Shao et al., "Projections of the Cost of Cancer Care in the United States: 2010–2020," *Journal of the National Cancer Institute* 103 (2011): 117–28. See also M. J. Hassett and E. B. Elkin, "What Does Breast Cancer Treatment Cost and What Is It Worth?," *Hematology/Oncology Clinics of North America* 27 (2013): 829–41.

155 In 2008, the national Coping with Cancer project: A. A. Wright et al., "Associations Between End-of-Life Discussions, Patient Mental Health, Medical Care Near Death, and Caregiver Bereavement Adjustment," *Journal of the American Medical Association* 300 (2008): 1665–73.

155 People with serious illness have priorities: P. A. Singer, D. K. Martin, and M. Kelner, "Quality End-of-Life Care: Patients' Perspectives," *Journal of the American Medical Association* 281 (1999): 163–68; K. E. Steinhauser et al., "Factors Considered Important at the End of Life by Patients, Family, Physicians, and Other Care Providers," *Journal of the American Medical Association* 284 (2000): 2476.

156 But as end-of-life researcher Joanne Lynn: J. Lynn, *Sick to Death and Not Going to Take It Anymore* (University of California Press, 2004).

156 Guides to *ars moriendi*: J. Shinners, ed., *Medieval Popular Religion 1000–1500: A Reader*, 2nd ed. (Broadview Press, 2007).

156 Last words: D. G. Faust, *This Republic of Suffering* (Knopf, 2008), pp. 10–11.

156 swift catastrophic illness is the exception: M. Heron, "Deaths: Leading Causes for 2009," National Vital Statistics Reports 61 (2009), http://www.cdc.gov/nchs/data/nvsr/nvsr61/nvsr61_07.pdf. See also Organisation for Economic Cooperation and Development, *Health at a Glance 2013*, http://www.oecd.org/els/health-systems/health-at-a-glance.htm.

167 First, our own views may be unrealistic: N. A. Christakis and E. B. Lamont, "Extent and Determinants of Error in Doctors' Prognoses in Terminally Ill Patients: Prospective Cohort Study," *BMJ* 320 (2000): 469–73.

167 Second, we often avoid voicing: E. J. Gordon and C. K. Daugherty, "'Hitting You Over the Head': Oncologists' Disclosure of Prognosis to Advanced Cancer Patients," *Bioethics* 17 (2003): 142–68; W. F. Baile et al., "Oncologists' Attitudes Toward and Practices in Giving Bad

News: An Exploratory Study," *Journal of Clinical Oncology* 20 (2002): 2189–96.

170 Gould published an extraordinary essay: S. J. Gould, "The Median Isn't the Message," *Discover*, June 1985.

174 the case of Nelene Fox: R. A. Rettig, P. D. Jacobson, C. Farquhar, and W. M. Aubry, *False Hope: Bone Marrow Transplantation for Breast Cancer* (Oxford University Press, 2007).

175 Ten states enacted laws: Centers for Diseases Control, "State Laws Relating to Breast Cancer," 2000.

175 Never mind that Health Net was right: E. A. Stadtmauer, A. O'Neill, L. J. Goldstein et al., "Conventional-Dose Chemotherapy Compared with High-Dose Chemotherapy plus Autologous Hematopoietic Stem-Cell Transplantation for Metastatic Breast Cancer," *New England Journal of Medicine* 342 (2000): 1069–76. See also Rettig et al., *False Hope*.

175 Aetna, decided to try a different approach: R. Krakauer et al., "Opportunities to Improve the Quality of Care for Advanced Illness," *Health Affairs* 28 (2009): 1357–59.

176 A two-year study of this "concurrent care" program: C. M. Spettell et al., "A Comprehensive Case Management Program to Improve Palliative Care," *Journal of Palliative Medicine* 12 (2009): 827–32. See also Krakauer et al. "Opportunities to Improve."

176 Aetna ran a more modest concurrent care program: Spettel et al., "A Comprehensive Case Management Program."

177 Two-thirds of the terminal cancer patients: Wright et al., "Associations Between End-of-Life Discussions."

177 A landmark 2010 study from the Massachusetts General Hospital: J. S. Temel et al., "Early Palliative Care for Patients with Metastatic Non-Small Cell Lung Cancer," *New England Journal of Medicine* 363 (2010): 733–42; J. A. Greer et al., "Effect of Early Palliative Care on Chemotherapy Use and End-of-Life Care in Patients with Metastatic Non-Small Cell Lung Cancer," *Journal of Clinical Oncology* 30 (2012): 394–400.

178 In one, researchers followed 4,493 Medicare patients: S. R. Connor et al., "Comparing Hospice and Nonhospice Survival among Patients Who Die Within a Three-Year Window," *Journal of Pain and Symptom Management* 33 (2007): 238–46.

179 By 1996, 85 percent of La Crosse residents: B. J. Hammes, *Having Your Own Say: Getting the Right Care When It Matters Most* (CHT Press, 2012).

192 Five of the ten fastest-growing: Data analyzed from World Bank, 2013, http://www.worldbank.org/en/publication/global-economic-prospects.

192 By 2030, one-half to two-thirds: Ernst & Young, "Hitting the Sweet Spot: The Growth of the Middle Class in Emerging Markets," 2013.

192 Surveys in some African cities: J. M. Lazenby and J. Olshevski, "Place of Death among Botswana's Oldest Old," *Omega* 65 (2012): 173–87.

192 leading families to empty bank accounts: K. Hanson and P. Berman, "Private Health Care Provision in Developing Countries: A Preliminary Analysis of Levels and Composition," *Data for Decision Making Project* (Harvard School of Public Health, 2013), http://www .hsph.harvard.edu/ihsg/topic.html.

192 Yet at the same time, hospice programs are appearing everywhere: H. Ddungu, "Palliative Care: What Approaches Are Suitable in the Developing World?," *British Journal of Haemotology* 154 (2011): 728–35. See also D. Clark et al., "Hospice and Palliative Care Development in Africa," *Journal of Pain and Symptom Management* 33 (2007): 698–710; R. H. Blank, "End of Life Decision-Making Across Cultures," *Journal of Law, Medicine & Ethics* (Summer 2011): 201–14.

192 Scholars have posited: D. Gu, G. Liu, D. A. Vlosky, and Z. Yi, "Factors Associated with Place of Death Among the Oldest Old," *Journal of Applied Gerontology* 26 (2007): 34–57.

193 Use of hospice care has been growing: National Center for Health Statistics, "Health, United States, 2010: With Special Feature on Death and Dying," 2011. See also National Hospice and Palliative Care Organization, "NHPCO Facts and Figures: Hospice Care in America, 2012 Edition," 2012.

198 Patients tend to be optimists: J. C. Weeks et al., "Patients' Expectations about Effects of Chemotherapy for Advanced Cancer," *New England Journal of Medicine* 367 (2012): 1616–25.

199 a short paper by two medical ethicists: E. J. Emanuel and L. L. Emanuel, "Four Models of the Physician-Patient Relationship," *Journal of the American Medical Association* 267 (1992): 2221–26.

203 most ovarian cancer patients at her stage: "Ovarian Cancer," online American Cancer Society guide, 2014, http://www.cancer.org/cancer /ovariancancer/detailedguide.

206 Bob Arnold, a palliative care physician I'd met: See A. Back, R. Arnold, and J. Tulsky, *Mastering Communication with Seriously Ill Patients* (Cambridge University Press, 2009).

223 One-third of the county lived in poverty: Office of Research, Ohio Development Services Agency, *The Ohio Poverty Report, February 2014* (ODSA, 2014), http://www.development.ohio.gov/files/research /P7005.pdf.

224 they formed Athens Village on the same model: More information at http://www.theathensvillage.org. They could use your donations, by the way.

8: COURAGE

231 Plato wrote a dialogue: *Laches*, trans. Benjamin Jowett, 1892, available online through Perseus Digital Library, Tufts University, http:// www.perseus.tufts.edu/hopper/text?doc=Perseus%3atext%3a1999 .01.0176%3atext%3dLach.

236 The brain gives us two ways to evaluate experiences: D. Kahneman, *Thinking, Fast and Slow* (Farrar, Straus, and Giroux, 2011). See also D. A. Redelmeier and D. Kahneman, "Patients' Memories of Painful Treatments: Real-Time and Retrospective Evaluations of Two Minimally Invasive Procedures," *Pain* 66 (1996): 3–8.

239 "An inconsistency is built into the design of our minds": Kahneman, *Thinking, Fast and Slow*, p. 385.

243 After some resistance, cardiologists now accept: A. E. Epstein et al., "ACC/AHA/HRS 2008 Guidelines for Device-Based Therapy of Cardiac Rhythm Abnormalities," *Circulation* 117 (2008): e350–e408. See also R. A. Zellner, M. P. Aulisio, and W. R. Lewis, "Should Implantable Cardioverter-Defibrillators and Permanent Pacemakers in Patients with Terminal Illness Be Deactivated? Patient Autonomy Is Paramount," *Circulation: Arrhythmia and Electrophysiology* 2 (2009): 340–44.

244 only a minority of people saved from suicide make a repeated attempt: S. Gibb et al. "Mortality and Further Suicidal Behaviour After an Index Suicide Attempt: A 10-Year Study," *Australia and New Zealand Journal of Psychiatry* 39 (2005): 95–100.

244 In places that allow physicians to write lethal prescriptions: E.g., the state of Washington's Death with Dignity Act, http://apps.leg.wa.gov /rcw/default.aspx?cite=70.245.

245 one in thirty-five Dutch people: Netherlands Government, "Euthanasia Carried Out in Nearly 3 Percent of Cases," *Statistics Netherlands*, July 21, 2012, http://www.cbs.nl/en-GB/menu/themas/gezondheid -welzijn/publicaties/artikelen/archief/2012/2012-3648-wm.htm.

245 The Dutch have been slower: British Medical Association, *Euthana-*

sia: Report of the Working Party to Review the British Medical Association's Guidance on Euthanasia, May 5, 1988, p. 49, n. 195. See also A.-M. The, *Verlossers Naast God: Dokters en Euthanasie in Nederland* (Thoeris, 2009).

245 About half don't even use their prescription: E.g., data from Oregon Health Authority, *Oregon's Death with Dignity Act, 2013 Report*, http://public.health.oregon.gov/ProviderPartnerResources/Evalua tionResearch/DeathwithDignityAct/Documents/year16.pdf.

249 Technological society has forgotten: L. Emanuel and K. G. Scandrett, "Decisions at the End of Life: Have We Come of Age?," *BMC Medicine* 8 (2010): 57.

Acknowledgments

I have a lot of people to thank for this book. First and foremost are my mother, Sushila Gawande, and my sister, Meeta. In choosing to include the story of my father's decline and death, I know I dredged up moments they'd rather not relive or necessarily have told the way I did. Nonetheless, they helped me at every turn, answering my difficult questions, probing their memories, and tracking down everything from memorabilia to medical records.

Other relatives here and abroad provided essential assistance as well. In India, my uncle Yadaorao Raut in particular sent me old letters and photographs, gathered memories about my father and grandfather from family members, and helped me check numerous details. Nan, Jim, Chuck, and Ann Hobson were equally generous with their memories and records of Alice Hobson's life.

I am also indebted to the many people I got to know and interview about their experiences with aging or serious illness, or dealing with a family member's. More than two hundred people gave me their time, told me their stories, and let me see into their lives. Only a fraction of them are explicitly mentioned in these pages. But they are all here just the same.

There were also scores of frontline staff in homes for the aged, palliative care experts, hospice workers, nursing home reformers, pioneers, and contrarians who showed me places and ideas I'd never have encountered. I especially want to thank two people: Robert Jenkens opened doors and guided me through the large community of people who are reinventing support for the aged, and Susan Block of the Dana Farber Cancer Institute not only did the same for the world of palliative and hospice care but also let me become her partner in research into how we might make the insights described here part of the fabric of care where we work and beyond it.

The Brigham and Women's Hospital and the Harvard School of Public Health have given me an incredible home for my work for more than a decade and a half. And my team at Ariadne Labs, the joint innovation center that I lead, has made mixing surgery, health systems research, and writing not only feasible but also a joy. This book would not have been possible without the particular efforts of Khaleel Seecharan, Katie Hurley, Kristina Vitek, Tanya Palit, Jennifer Nadelson, Bill Berry, Arnie Epstein, Chip Moore, and Michael Zinner. Dalia Littman helped with fact-checking. And, most indispensably, the brilliant and unfazable Ami Karlage spent the last three years working on this book with me as research assistant, storyboard artist, manuscript organizer, sounding board, and supplier, when necessary, of Bourbon Brambles.

The *New Yorker* magazine has been my other creative home. I count myself as unfairly lucky not only to have gotten to write for that amazing publication (thank you, David Remnick) but also to have had as my editor and friend the great Henry Finder. He saw me through writing the two essays for the magazine that became the foundation of this book and guided me to many pivotal additional ideas. (He was, for instance, the one who told me to read Josiah Royce.)

Tina Bennett has been my tireless agent, my unstinting protector, and, going all the way back to college, my dear friend. Although everything about publishing books is changing, she has always found a way for me to grow an audience and still write what I want to write. She is without peer.

The Rockefeller Foundation provided its gorgeous Bellagio Center as a retreat where I started the book and then returned to finish the first draft. My subsequent conversations about that manuscript with Henry, Tina, David Segal, and Jacob Weisberg transformed the way I saw the book, leading me to remake it from beginning to end. Leo Carey did a line edit of the final draft, and his ear for language and clear expression made the book tremendously better. Riva Hocherman helped greatly at every stage, including providing an invaluable final read through. Thank you also to Grigory Tovbis and Roslyn Schloss for their essential contributions.

My wife, Kathleen Hobson, has been more important to this book than she knows. Every idea and story here we have talked through together and in many instances also lived through together. She has been a constant, encouraging force. I have never been a facile writer. I don't know what those authors who describe the words just flowing out of them are talking about. For me, the words come only slowly and after repeated effort. But Kathleen has always helped me find the words and made me know the work is achievable and worthwhile no matter how long it takes. She and our three amazing children, Hunter, Hattie, and Walker, have pulled for me each step of the way.

Then there is Sara Bershtel, my extraordinary editor. As she worked on the book, Sara was forced to live through its most difficult realities in her own family. It would have been understandable for her to choose to step aside. But her devotion to the book remained unwavering, and she went through every draft with me meticulously, working paragraph by paragraph to make sure I'd got every part as true and right as I could. Sara's dedication is the reason this book says what I wanted it to say. And that is why it is dedicated to her.